Limit Theorems of Polynomial Approximation with Exponential Weights

of the
American Mathematical Society

Number 897

Limit Theorems of Polynomial Approximation with Exponential Weights

Michael I. Ganzburg

March 2008 • Volume 192 • Number 897 (second of 5 numbers) • ISSN 0065-9266

American Mathematical Society
Providence, Rhode Island

2000 *Mathematics Subject Classification.* Primary 30D15, 41A30; Secondary 31A05, 41A17.

Library of Congress Cataloging-in-Publication Data

Ganzburg, Michael I., 1948–

Limit theorems of polynomial approximation with exponential weights / Michael I. Ganzburg.

p. cm. — (Memoirs of the American Mathematical Society, ISSN 0065-9266 ; no. 897)

Includes bibliographical references and index.

ISBN 978-0-8218-4063-4 (alk. paper)

1. Functions, Entire. 2. Approximation theory. 3. Potential theory (Mathematics) 4. Fourier analysis.

QA353.E5G36   2008
510 s—dc22
[515′.98]
                                                                              2007060582

Memoirs of the American Mathematical Society

This journal is devoted entirely to research in pure and applied mathematics.

Subscription information. The 2008 subscription begins with volume 191 and consists of six mailings, each containing one or more numbers. Subscription prices for 2008 are US$675 list, US$540 institutional member. A late charge of 10% of the subscription price will be imposed on orders received from nonmembers after January 1 of the subscription year. Subscribers outside the United States and India must pay a postage surcharge of US$38; subscribers in India must pay a postage surcharge of US$43. Expedited delivery to destinations in North America US$53; elsewhere US$130. Each number may be ordered separately; *please specify number* when ordering an individual number. For prices and titles of recently released numbers, see the New Publications sections of the *Notices of the American Mathematical Society*.

Back number information. For back issues see the *AMS Catalog of Publications*.

Subscriptions and orders should be addressed to the American Mathematical Society, P. O. Box 845904, Boston, MA 02284-5904, USA. *All orders must be accompanied by payment.* Other correspondence should be addressed to 201 Charles Street, Providence, RI 02904-2294, USA.

Copying and reprinting. Individual readers of this publication, and nonprofit libraries acting for them, are permitted to make fair use of the material, such as to copy a chapter for use in teaching or research. Permission is granted to quote brief passages from this publication in reviews, provided the customary acknowledgement of the source is given.

Republication, systematic copying, or multiple reproduction of any material in this publication is permitted only under license from the American Mathematical Society. Requests for such permission should be addressed to the Acquisitions Department, American Mathematical Society, 201 Charles Street, Providence, Rhode Island 02904-2294, USA. Requests can also be made by e-mail to reprint-permission@ams.org.

Memoirs of the American Mathematical Society is published bimonthly (each volume consisting usually of more than one number) by the American Mathematical Society at 201 Charles Street, Providence, RI 02904-2294, USA. Periodicals postage paid at Providence, RI. Postmaster: Send address changes to Memoirs, American Mathematical Society, 201 Charles Street, Providence, RI 02904-2294, USA.

© 2008 by the American Mathematical Society. All rights reserved.
Copyright of this publication reverts to the public domain 28 years after publication. Contact the AMS for copyright status.
This publication is indexed in *Science Citation Index*®, *SciSearch*®, *Research Alert*®, *CompuMath Citation Index*®, *Current Contents*®/*Physical, Chemical & Earth Sciences*.
Printed in the United States of America.

∞ The paper used in this book is acid-free and falls within the guidelines established to ensure permanence and durability.
Visit the AMS home page at http://www.ams.org/

10 9 8 7 6 5 4 3 2 1   13 12 11 10 09 08

Contents

Chapter 1. Introduction 1
  1.1. A Brief Review 1
  1.2. Results and Organization of the Monograph 5
  1.3. Basic Notation and Some Preliminaries 7
  1.4. Classes of Weights and Basic Estimates 8
  1.5. Acknowledgements 13

Chapter 2. Statement of Main Results 15
  2.1. Limit Theorems of Polynomial Approximation with Exponential Weights 15
  2.2. Approximation of Entire Functions of Exponential Type 16
  2.3. Polynomial Inequalities in the Complex Plane 17

Chapter 3. Properties of Harmonic Functions 19
  3.1. The Poisson Integral Re $H(w)$ 19
  3.2. The Function $h(r)$ and the Constant b_n 24
  3.3. The Functions $\phi(r)$ and $\phi_1(r)$ 29
  3.4. The Main Estimate for Re $H(w)$ 35

Chapter 4. Polynomial Inequalities with Exponential Weights 43
  4.1. Nikolskii-type Inequalities 43
  4.2. Extremal Polynomials 45
  4.3. Polynomial Inequalities in the Complex Plane 55
  4.4. Proofs of Theorems 2.3.1 and 2.3.2 57

Chapter 5. Entire Functions of Exponential Type and their Approximation Properties 59
  5.1. Entire Functions of Exponential Type 59
  5.2. Approximation Properties of Entire Functions of Exponential Type 62

Chapter 6. Polynomial Interpolation and Approximation of Entire Functions of Exponential Type 67
  6.1. Interpolation on the Interval $I_n = [-a_n(1+\delta_n), a_n(1+\delta_n)]$ 67
  6.2. Interpolation on $I \setminus I_n$ 71
  6.3. Proof of Theorem 2.2.1 72
  6.4. Proof of Theorem 2.2.2 74

Chapter 7. Proofs of the Limit Theorems 77
  7.1. Proof of Theorem 2.1.1 77
  7.2. Proof of Theorem 2.1.2 80
  7.3. Proofs of Theorems 2.1.3 and 2.1.4 82

Chapter 8. Applications   85
  8.1. Approximation of Individual Functions and Proof of Theorem 2.3.3   85
  8.2. An Asymptotically Sharp Constant of Weighted Approximation on
      the Class $W^r H^\lambda[I]$ 96
  8.3. Convergence of Polynomials and a Mehler-Heine Formula for
      Orthonormal Polynomials   100

Chapter 9. Multidimensional Limit Theorems of Polynomial Approximation
        with Exponential Weights   105
  9.1. Multidimensional Limit Theorems with Exponential Weights   105
  9.2. Proof of Theorem 9.1.3   108
  9.3. Proofs of Theorems 9.1.1 and 9.1.4   111
  9.4. Approximation of λ-Homogeneous Functions   113

Chapter 10. Examples   117
  10.1. $W(x) = \exp(-|x|^\alpha)$, $\alpha > 1$ 117
  10.2. $W(x) = \exp(-|x|)$ 121
  10.3. $W(x) = \exp(-|x|^\alpha)$, $0 < \alpha < 1$ 127
  10.4. $W(x) = \exp(-|x|^\alpha)$, $\alpha \to \infty$ 132
  10.5. Examples of Erdös Weights   134

Appendix A. Appendix. Negativity of a Kernel   137
  A.1. Statement of the Main Result   137
  A.2. Some Technical Results   138
  A.3. Proof of Proposition A.1.1   144

Bibliography   155

Index   161

Abstract

We develop the limit relations between the errors of polynomial approximation in weighted metrics and apply them to various problems in approximation theory such as asymptotically best constants, convergence of polynomials, approximation of individual functions, and multidimensional limit theorems of polynomial approximation.

In addition we establish new inequalities for polynomials in complex domains and new asymptotics and estimates for orthogonal polynomials with exponential weights. More detailed information on approximation properties of functions is obtained for the canonical weights $W(x) = \exp(-|x|^\alpha)$, $0 < \alpha < \infty$, and some other weights.

Received by the editor December 13, 2004.

2000 *Mathematics Subject Classification*. Primary 30D15, 41A30; Secondary 31A05, 41A17, 41A44, 42C05.

Key words and phrases. Polynomial approximation, exponential weights, entire functions of exponential type, polynomial inequalities, harmonic functions.

CHAPTER 1

Introduction

1.1. A Brief Review

In this monograph, we discuss several topics of analysis such as polynomial approximation and orthogonal polynomials with exponential weights, approximation by entire functions of exponential type, polynomial inequalities, and harmonic functions. The focus of the work is on the limit relations between the errors of polynomial approximation in weighted metrics and on their applications in approximation theory.

In May 1923, Bernstein [8, p. 399] (see also [9, 10]) initiated the study of polynomial approximation on $(-\infty, \infty)$ with an exponential weight $W(x) = \exp(-Q(x))$ by posing a problem that has attracted much attention for more than 30 years. Bernstein's approximation problem was to determine the necessary and sufficient conditions on W such that the set $\mathcal{P}$ of all polynomials is dense in the weighted space $L^0_{\infty,W}(\mathbb{R})$ of all continuous functions f with the norm $\|f\|_{L_{\infty,W}(\mathbb{R})} = \sup_{x \in \mathbb{R}} |f(x)|W(x)$, satisfying the condition $\lim_{|x| \to \infty} f(x)W(x) = 0$.

The full solution to this problem was given in the 1950s independently by Akhiezer and Bernstein [5], Mergelyan [87], and Pollard [103] (for a continuous W), see surveys [3, 88] and books [4, 59, 74] for detailed discussions on these developments and for further references.

The most elegant solution to Bernstein's problem was found for certain regular weights.

THEOREM 1.1.1. (Dzrbasjan [28], Carleson [59]). *If Q is an even continuous function on $(-\infty, \infty)$ and $Q(e^x)$ is convex, then $\mathcal{P}$ is dense in $L^0_{\infty,W}(\mathbb{R})$ if and only if*
$$\int_{-\infty}^{\infty} \frac{Q(t)}{1+t^2} \, dt = \infty.$$

In particular, for $W_\alpha(x) := \exp(-|x|^\alpha)$, $\alpha > 0$, it easily follows from Theorem 1.1.1 that $\mathcal{P}$ is dense in $L^0_{\infty,W_\alpha}(\mathbb{R})$ if and only if $\alpha \geq 1$.

The first results in quantitative polynomial approximation with exponential weights were obtained in the 1950s by Dzrbasjan and his students. In particular, Dzrbasjan [29] found the correct rate of approximation of $|x|$ in the weighted uniform metric.

It was Freud who in the 1960s–1970s began using orthogonal polynomials as the main tool in the study of weighted approximation. Applying Lagrange interpolation, orthogonal expansions, estimates for Cristoffel functions, and other techniques to weighted approximation, Freud [36, 37] proved Markov-Bernstein inequalities and established direct and converse theorems for approximation with even exponential weights of smooth polynomial growth at ∞ (Freud weights).

One more piece of the puzzle was added by Mhaskar and Saff [93–95] in the mid-1980s. Using elementary potential theory, these authors showed that for an even weight $W(x) = \exp(-Q(x))$ with a strictly increasing $xQ'(x)$ and for any polynomial P of degree at most n, the function PW lives on the interval $[-a_n, a_n]$, i.e.

(1.1.1) $$\|PW\|_{L_\infty(\mathbb{R})} = \|PW\|_{L_\infty[-a_n, a_n]}.$$

Here, a_n is the nth Mhaskar-Rakhmanov-Saff number defined as the positive root of the equation
$$n = \frac{2}{\pi} \int_0^1 \frac{a_n x Q'(a_n x)}{\sqrt{1-x^2}} dx.$$
Note that Rakhmanov [107] independently introduced this number using potential theory in his study of orthogonal polynomials. Since then L_p-analogues of (1.1.1), $0 < p \leq \infty$ (so-called infinite-finite range inequalities) and potential theory became the most powerful tools in weighted approximation and orthogonal polynomials (see [72, 74, 91, 109]).

The parallel development of weighted approximation and orthogonal polynomials on $(-\infty, \infty)$ has influenced over intense research in both areas.

Extension of known results to more general weights and classes of functions and finding more precise estimates and asymptotics has been the main direction of research in the late 1980s–1990s. In particular, Rakhmanov [108] found the strong asymptotics for the orthogonal polynomials associated with the weight W_α^2, $\alpha > 1$. A nontrivial extension of these results to general Freud weights and to more general classes of polynomials was obtained by Lubinsky and Saff [84]. Similar results for exponential weights $W(x) = \exp(-Q(x))$ with Q, having faster than polynomial growth on ∞ (Erdös weights), was proved by Lubinsky [75]. Totik [116] gave a simplified proof of these asymptotics and extended them to more general weights. Deift, Kriecherbauer, McLaughlin, Venakides, and Zhou [23, 64] used the Riemann-Hilbert techniques to prove far more precise asymptotics for the canonical weights W_α, $\alpha > 0$.

In weighted approximation, Markov-Bernstein inequalities for Freud and Erdös weights were established in $L_p(\mathbb{R})$-metric, $0 < p \leq \infty$, in both uniform and pointwise forms by Nevai and Totik [98], Lubinsky and Nevai [83], Levin and Lubinsky [69, 70], Lubinsky [76], Lubinsky and Mthembu [82], Mthembu [96]. The direct and converse theorems of polynomial approximation with Freud weights in $L_p(\mathbb{R})$, $1 \leq p \leq \infty$, were proved by Ditzian and Totik in [27] (see also [91]). Ditzian and Lubinsky [26] extended these results to $p \in (0, 1)$. For Erdös weights, the corresponding theorems were established by Damelin and Lubinsky [22] and Damelin [21].

Detailed discussions on the development of weighted approximation and orthogonal polynomials on $(-\infty, \infty)$ and further references can be found in surveys [77, 78, 80, 97] and in books [26, 72, 91, 109, 116].

The asymptotic behavior of orthogonal polynomials associated with a weight W on a bounded interval (say, $[-1, 1]$) was found by Szegö [113] if

(1.1.2) $$\int_{-1}^1 \frac{\log W(x)}{\sqrt{1-x^2}} dx > -\infty.$$

The case when condition (1.1.2) is violated has been discussed in many papers (see [110, 118]). In particular, Levin and Lubinsky [71] studied Christoffel functions and Markov-Bernstein inequalities for some non-Szegö weights.

A unified approach to orthogonal polynomials and to Markov-Bernstein inequalities with exponential weights on bounded or unbounded intervals was developed in the recent monograph of Levin and Lubinsky [72].

The techniques developed in the 1970s–1990s were sufficient to find constructive characteristics of fairly general weighted classes of functions on infinite interval (cf. [21, 22, 26, 27, 91]). The corresponding results for finite intervals can be established similarly (cf. [79]). However, in order to solve such problems as finding the asymptotic behavior of the weighted approximation error of individual functions or classes of functions, a more delicate approach is needed.

In the 1940s, Bernstein [13, 16] came up with such an approach for nonweighted approximation (the limit theorems for polynomial approximation).

Let $\mathcal{P}_n$ be the class of algebraic polynomials of degree at most n and B_σ the class of all entire functions of exponential type $\sigma > 0$. Let us denote by

$$E_n(f, L_p(\Omega)) := \inf_{P \in \mathcal{P}_n} \|f - P\|_{L_p(\Omega)}, \quad A_\sigma(f, L_p) := \inf_{g \in B_\sigma} \|f - g\|_{L_p(\mathbb{R})},$$

the errors of approximation of a measurable f by polynomials and entire functions of exponential type. Here, $0 < p \leq \infty$ and $\Omega \subseteq \mathbb{R}$ is either an interval or the real line $\mathbb{R}$.

Bernstein [13, 114, p. 48] proved that the relation

$$(1.1.3) \qquad \lim_{n \to \infty} E_n(f, L_\infty[-n/\sigma, n/\sigma]) = A_\sigma(f, L_\infty)$$

holds for each continuous function of polynomial growth on $\mathbb{R}$ and for almost all $\sigma > 0$. It was shown by the author [44] that (1.1.3) is invalid for $f(x) = \cos x$ and $\sigma = 1$. However, the following equality holds for all $\sigma > 0$ [44]:

$$(1.1.4) \qquad \lim_{n \to \infty} E_n\left(f, L_\infty\left[-\frac{n(1-\lambda_n)}{\sigma}, \frac{n(1-\lambda_n)}{\sigma}\right]\right) = A_\sigma(f, L_\infty),$$

where $\{\lambda_n\}_{n=1}^\infty$ is a sequence of real numbers satisfying the conditions:
  (a) $0 \leq \lambda_n \leq 1$, $n = 1, 2, \ldots$,
  (b) $\lambda_n = o(1)$, $n \to \infty$,
  (c) for some $\delta \in (0, 2/3)$, $\liminf_{n \to \infty} \lambda_n n^\delta > 0$.

Note that these conditions are unimprovable in a certain sense [44].

It is easy to see that (1.1.4) is equivalent to the relation

$$(1.1.5) \qquad \lim_{n \to \infty} E_n\left(f\left(\frac{n(1-\lambda_n)}{\sigma} \cdot\right), L_\infty[-1, 1]\right) = A_\sigma(f, L_\infty).$$

For $0 < p < \infty$, $\sigma > 0$ and for a measurable f of polynomial growth with $A_\sigma(f, L_p) < \infty$, the following L_p-analogue of (1.1.5) holds for $\lambda_n = 0$:

$$(1.1.6) \qquad \lim_{n \to \infty} n^{1/p} E_n(f((n/\sigma)\cdot), L_p[-1, 1]) = \sigma^{1/p} A_\sigma(f, L_p).$$

For $p \in [1, \infty)$ this result was announced by Bernstein [16] and proved by Raitsin [105]. The author established (1.1.6) for $p \in (0, 1)$ in [47].

It is possible to extend (1.1.5) and (1.1.6) to any function (cf. [16, 44, 4, p. 368, 114, p. 286]). Namely, let f be continuous on $\mathbb{R}$ if $p = \infty$ and let f be measurable on $\mathbb{R}$ if $0 < p < \infty$. If for some $\sigma_0 > 0$, $A_{\sigma_0}(f, L_p) < \infty$, then for any $\sigma > \gamma\sigma_0$ relation (1.1.5) holds if $p = \infty$ and (1.1.6) holds if $0 < p < \infty$. Here

(1.1.7) $$\gamma := 1.508879\ldots$$

is the positive solution to the equation

(1.1.8) $$\sqrt{\tau^2 + 1}/\tau = \log(\sqrt{\tau^2 + 1} + \tau).$$

Note that in this statement γ cannot be replaced by a lesser number (cf. [16, 4, p. 365]).

Various generalizations of limit theorems (1.1.5) and (1.1.6) to the multidimensional case were obtained in [40, 43, 44]. A general approach to the limit theorems was developed in [45, 47].

Numerous applications of the limit theorems to Bernstein-Nikolskii inequalities [38, 41, 48], Jackson-type theorems [38, 39, 114, p. 257], sharp constants of approximation theory [15, 40, 42, 47, 51], approximation of individual functions [14, 40, 49, 4, p. 371, 114, p. 416] and properties of the approximation errors [43] show that these theorems can be applied to various problems of approximation theory which cannot be solved by using other methods. Also, the use of the limit theorems essentially reduces the proof of some known results. For example, the celebrated theorems of Bernstein [11] and Nikolskii [99] on the asymptotic behavior of $E_n(|x|^\lambda, L_\infty[-1,1])$ and $E_n(|x|^\lambda, L_1[-1,1])$ immediately follows from (1.1.5) and (1.1.6), see Section 8.1 for details.

The proofs of (1.1.5) and (1.1.6) and a number of applications of the limit theorems are based on an estimate [44]

(1.1.9) $$E_n\left(g\left(\frac{n(1-\lambda_n)}{\sigma}\cdot\right), L_p[-1,1]\right) \leq C_1 \|g_\sigma\|_{L_\infty(\mathbb{R})} \exp(-C_2 n^\delta),$$

where g is a bounded entire function of exponential type σ. The similar estimates hold for $g \in B_\sigma$ of polynomial growth on $\mathbb{R}$ and for any $g \in B_\sigma$, $\sigma > 0$ [44].

Relations like (1.1.5), (1.1.6), and (1.1.9) serve as the "bridges" between polynomial and harmonic approximations and make it possible to transfer results from one area of approximation theory to another. For example, the V. A. Markov inequality for $P \in \mathcal{P}_n$ (see [114, p. 226]),

(1.1.10) $$|P'(0)| \leq (n/c)\|P\|_{L_\infty[-c,c]},$$

combined with (1.1.9), imply the Bernstein inequality for the derivative of a bounded $g \in B_\sigma$, $\sigma > 0$ [114, p. 208]. Indeed, it follows from (1.1.9) that

(1.1.11) $$\|g - P_n\|_{L_\infty[-c_n,c_n]} + \|g' - P_n'\|_{L_\infty[-c_n,c_n]} \leq C_3 \|g\|_{L_\infty(\mathbb{R})} \exp(-C_4 n^{\delta_1}),$$

where $\delta_1 > 0$, $c_n := n(1-\lambda_n)/\sigma$, and P_n is the polynomial of best approximation to g in $L_\infty[-c_n, c_n]$, $n = 1, 2, \ldots$.

Then by (1.1.10) and (1.1.11), we have

$$|g'(0)| \leq \liminf_{n\to\infty}\left(|P_n'(0)| + \|g' - P_n'\|_{L_\infty[-c_n,c_n]}\right) \leq \sigma\|g\|_{L_\infty(\mathbb{R})}.$$

Hence the Bernstein inequality follows. Note that a general approach to Bernstein-type inequalities was developed in [41].

Also, applying Jackson's theorem for polynomial approximation [25, p. 220] to the left-hand sides of (1.1.5) and (1.1.6), we arrive at the corresponding estimates for $A_\sigma(f, L_p)$ (cf. [114, p. 257]). On the other hand, Jackson-type estimates for polynomial approximation discussed in [39] are based on Jackson's theorem for approximation by entire functions of exponential type and on estimates like (1.1.9).

The development of the limit theorems and their applications in approximation theory were discussed in surveys [46, 47].

1.2. Results and Organization of the Monograph

In this monograph, we discuss the limit theorems like (1.1.5), (1.1.6) and estimates like (1.1.9) with exponential weights on bounded or unbounded intervals $(-c, c)$ and study their applications in weighted approximation. These results are proved for the wide class $\mathcal{F}(C^2)$ of even smooth weights introduced in [72] that includes Freud and Erdös weights on $(-c, c)$. In particular, this class contains the canonical weights W_α, $\alpha > 1$.

We establish the basic limit relations for $W(x) = \exp(-Q(x)) \in \mathcal{F}(C^2)$ and for $p \in (0, \infty]$ of the form

$$\lim_{n \to \infty} (b_n/\sigma)^{1/p} E_n(f(\gamma_n \cdot), L_{p,W}(-c,c))$$
(1.2.1)
$$= \lim_{n \to \infty} (b_n/\sigma)^{1/p} E_n(f(\gamma_n \cdot), L_{p,W}[-a_n, a_n]) = A_\sigma(f, L_p),$$

where

$$E(h, L_{p,w}(\Omega)) := \inf_{P \in \mathcal{P}_n} \left(\int_\Omega (|h(x) - P(x)|W(x))^p \, dx \right)^{1/p}, \qquad \Omega \subseteq \mathbb{R},$$

a_n is the Mhaskar-Rakhmanov-Saff number,

$$\gamma_n := \begin{cases} b_n(1 - O(n^{-\delta}))/\sigma, & p = \infty \\ b_n/\sigma, & 0 < p < \infty, \end{cases}$$

and b_n is the number defined below:

(1.2.2)
$$b_n = b_n(Q) := \frac{2}{\pi} \int_0^1 \frac{Q'(a_n y)\sqrt{1 - y^2}}{y} \, dy + \frac{n}{a_n}.$$

Note that $n/a_n \leq b_n \leq (1 + C)n/a_n$ and for Erdös weights, $b_n = (n/a_n)(1 + o(1))$ as $n \to \infty$.

Relations (1.2.1) are valid for a measurable (continuous if $p = \infty$) function f of polynomial growth on $\mathbb{R}$, provided that $A_\sigma(f, L_p) < \infty$. Extensions of (1.2.1) to all measurable (continuous) functions f are given as well. These results are proved in Chapter 7.

Two chief ingredients of the proofs that we believe to be of independent interest are a weighted version of estimate (1.1.9) and the polynomial inequality

(1.2.3)
$$|P(z)| \leq e^{b_n|z|} \|PW\|_{L_\infty(-c,c)}, \qquad P \in \mathcal{P}_n, \quad z \in \mathbb{C}.$$

These results are proved in Chapter 6 and Section 4.4, respectively, and based on weighted polynomial inequalities (Chapter 4) and on certain properties of harmonic functions (Chapter 3) and entire functions of exponential type (Chapter 5).

In particular, one of the basic constituents of the proof of (1.2.3) is the following inequality for the Poisson integral, proved in Section 3.4:

(1.2.4)
$$\frac{1-r^2}{2\pi} \int_{-\pi}^{\pi} \frac{Q(a_n \cos t)}{1 + r^2 - 2r\cos(t+\varphi)} \, dt \leq \frac{a_n b_n}{2} \left| w + \frac{1}{w} \right| + n \log |w|,$$

where $w = re^{i\varphi}$ and $0 < r \leq 1$.

The proof of a weighted version of (1.1.9) is essentially based on a lower estimate for the extremal polynomial $P_{n,q}(a_n u)$ associated with W, where $q \in (1, \infty)$ and $|u + \sqrt{u^2-1}| \geq 1 + O(n^{-\delta})$, $u \in \mathbb{C}$ (Section 4.2). For $|u + \sqrt{u^2-1}| \geq 1 + \varepsilon$, $u \in \mathbb{C}$, this estimate follows from the strong asymptotics for $P_{n,q}(a_n u)$ given in [72]. A weighted version of the V. A. Markov inequality for the coefficients of an algebraic polynomial is discussed in Section 4.4.

Applications of the limit theorems are given in Chapter 8. Section 8.1 contains the asymptotic formulae for the approximation error of λ-homogeneous functions $f_{\lambda, a, b}$ on $\mathbb{R}$ and functions of the form $f_{\lambda, a, b}(x) \log^k |x|$ in the weighted L_p-metrics. In particular, the following asymptotic holds:

$$(1.2.5) \qquad \lim_{n \to \infty} (b_n/\sigma)^{1/p} E_n(f_\lambda, L_{p,W}(-c, c)) = B_{\lambda, p}, \qquad 0 < p \leq \infty,$$

where $f_\lambda(x) := |x|^\lambda$, $\lambda > -1/p$, and $B_{\lambda,p}$ is the Bernstein constant (cf. [49]). Some estimates of $E_n(f_1, L_{\infty, W}(\mathbb{R}))$ were established earlier by Bernstein [10], Dzrbasjan [29], Kroó and Szabados [65]. Using (1.2.5), we also show that inequalities (1.2.3) and (1.2.4) are unimprovable in a certain sense.

In Section 8.2, we find an asymptotically sharp constant of weighted approximation on the class $W^r H^\alpha$. In the proof we use (1.2.1) and a recent solution [51] of the long-standing problem on the best constant of approximation by entire functions of exponential type on the class $W^r H_\omega$, where ω is a concave modulus of continuity. A Mehler-Heine type formula for orthogonal polynomials and its version for a sequence of polynomials are given in Section 8.3. Note that all theorems in Chapter 8 are similar to non-weighted results with n replaced by b_n.

Multidimensional limit theorems of polynomial approximation with exponential weights are discussed in Chapter 9.

Chapter 10 contains results on polynomial approximation with special exponential weights, mostly with the canonical weights W_α, $\alpha > 0$. In particular, in Section 10.1, we show that if $A_{\sigma_0}(f, L_p) < \infty$, then the second relation in (1.2.1) is valid for all $\sigma > \beta(\alpha)\sigma_0$, where

$$(1.2.6) \qquad \beta(\alpha) := \left[\frac{2(\alpha - 1)r_0(1 + r_0^2)}{\alpha(1 + r_0^2)^2 - (1 - r_0^2)^2} \log \frac{1}{r_0} \right]^{-1}$$

and $r_0 = r_0(\alpha)$ is the solution to the equation

$$(1.2.7) \qquad F(-\alpha/2, 1, \alpha/2 + 1, r^2) = \frac{1}{2} + \frac{(1 - r^2)^2}{2((1 + r^2)^2 - (1 - r^2)^2/\alpha)} \log \frac{1}{r}.$$

Note that letting $\alpha \to \infty$ in (1.2.6) and (1.2.7), we have $\lim_{\alpha \to \infty} \beta(\alpha) = \gamma$, where γ is given in (1.1.7). This is consistent with the result discussed in Section 10.4 that non-weighted polynomial approximation in $L_p[-1, 1]$ is the limit case of approximation with the weight W_α on $\mathbb{R}$ as $\alpha \to \infty$.

In Section 10.2, we consider the limit theorems for the boundary weight W_1. Theorem 1.1.1 shows that polynomials are still dense in L^0_{∞, W_1}, but this weight does not belong to $\mathcal{F}(C^2)$. In particular, $b_n(|\cdot|) = \infty$. That is why we need to reprove all basic technical results on polynomial inequalities in this weighted metric. The most significant technicalities used in this section are a version of (1.2.3) for this weight and a lower inequality for the orthonormal polynomials $P_{n,2}(a_n u)$ associated with W_1, where

(1.2.8) $$\left|u + \sqrt{u^2 - 1}\right| \geq 1 + Cn^{-1/3+\varepsilon}.$$

The latter result is proved in Section 4.2 and based on a very precise asymptotic for $P_{n,2}(a_n u)$ found by Kriecherbauer and McLaughlin [64]. However since the limit theorems for W_1 are very sensitive with respect to the exponent $1/3$ in (1.2.8), we are unable to prove limit relations like (1.2.1). Instead, we establish two-sided limit inequalities with b_n replaced by $2(1-\varepsilon)\log n/(3\pi\sigma)$ in the lower estimate and with b_n replaced by $2\log n/(\pi\sigma)$ in the upper one. The question what constant should replace b_n in (1.2.1) for $W = W_1$ is open and seems to be difficult.

In Section 10.3, we use techniques from Chapter 6 to describe the closure $\mathcal{P}[W_\alpha]$ of all polynomials in $L_{\infty,W_\alpha}(\mathbb{R})$, where $0 < \alpha < 1$. Theorem 1.1.1 shows that this class does not coincide with $L_{\infty,W_\alpha}(\mathbb{R})$. Moreover, Mergelyan [88] (see also [4, 59]) proved that every function from $\mathcal{P}[W_\alpha]$ is the restriction on $\mathbb{R}$ of an entire function of minimal type. Some descriptions of $\mathcal{P}[W_\alpha]$ are given by Kroó, Szabados, and Varga [66]. We give different two-sided inclusions for this set, though the problem of a complete description is open.

Some examples of Erdös weights on bounded and unbounded intervals are discussed in Section 10.5. In particular, the formulae for the constants a_n and b_n are given.

1.3. Basic Notation and Some Preliminaries

Throughout $C, C_1, C_2, \ldots$ denote positive constants independent of variables n, x, y, t, s, r, k, u, z, and polynomials $P, \ldots$, and functions $f, g, \ldots$. Occasionally we indicate dependence on, or independence of, certain parameters. The same symbol does not necessarily denote the same constant in different occurrences. We use the notation $g(x) \sim f(x)$ or $\alpha_n \sim \beta_n$ if there are positive constants C_1 and C_2 such that $C_1 \leq g(x)/f(x) \leq C_2$ for the relevant range of x or $C_1 \leq \alpha_n/\beta_n \leq C_2$ for all n large enough.

Let $0 < p \leq \infty$ and let $W(x) = \exp(-Q(x))$ be an exponential weight defined on an open symmetric interval I (bounded or unbounded) of the real axis $\mathbb{R}$. We assume that Q is a finite positive function on I, satisfying $Q(0) = 1$. Various classes of weights and some characteristics are defined in the next section.

Let $L_{p,W}(I)$ be the quasi-normed space of all measurable functions f on I with the finite quasi-norm

$$\|f\|_{L_{p,W}(I)} := \left(\int_I (|f(x)|w(x))^p \, dx\right)^{1/p}, \qquad 0 < p < \infty,$$

and let $L_{\infty,W}(I)$ be the space of all continuous functions f on I with the finite norm

$$\|f\|_{L_{\infty,W}(I)} := \sup_{x \in I} |f(x)|W(x).$$

In the non-weighted case ($W(x) = 1$), we set

$$\|\cdot\|_{L_p(I)} := \|\cdot\|_{L_{p,1}(I)}; \quad L_p(I) := L_{p,1}(I), \qquad 0 < p \leq \infty.$$

If E is a measurable subset of I, then $L_{p,W}(E)$ is a space of all f with the finite quasi-norm $\|f\|_{L_{p,W}(E)} := \|f\chi_E\|_{L_{p,W}(I)}$, where χ_E is the characteristic function of E, $0 < p \leq \infty$. In particular, for $E = [a, b]$ we set $L_{p,W}[a, b] := L_{p,W}([a, b])$. We also need the class $L_{p,\text{loc}}(\mathbb{R})$ of all measurable functions f on $\mathbb{R}$ such that $f \in L_p[-a,a]$ for any $a > 0$. The Lebesgue measure of a measurable set E is denoted by $|E|$.

For $n = 0, 1, \ldots$, $\mathcal{P}_n$ denotes the class of all algebraic polynomials of degree at most n.

Let B_σ, $\sigma > 0$, be the class of all entire functions of exponential type σ, i.e., the class of all entire functions g satisfying the condition: for any $\varepsilon > 0$ there exists a constant $C = C(g,\varepsilon)$ such that

$$|g(z)| \leq C e^{\sigma(1+\varepsilon)|z|}$$

for all points z from the complex plane $\mathbb{C}$.

Note that the restrictions of all functions and polynomials on $\mathbb{R}$ or on subsets of $\mathbb{R}$ are complex-valued unless indicated otherwise.

Let
$$\psi(z) := z + \sqrt{z^2 - 1}$$
denote the conformal map of $\mathbb{C} \setminus [-1,1]$ onto $\{z \in \mathbb{C} : |z| > 1\}$.

The Fourier transform and the inverse Fourier transform of a function or a tempered distribution f is denoted by $\hat{f}$ and $\check{f}$, respectively. In particular, for $f \in L_1(\mathbb{R})$,

$$\hat{f}(y) := (2\pi)^{-1/2} \int_{-\infty}^{\infty} f(x) e^{-ixy}\, dx; \quad \check{f}(y) := (2\pi)^{-1/2} \int_{-\infty}^{\infty} f(x) e^{ixy}\, dx.$$

For $0 < p \leq \infty$ and for a measurable (or continuous if $p = \infty$) function f on $\mathbb{R}$ or on an interval $\Omega \subseteq \mathbb{R}$, we define the approximation errors

$$E_n(f, L_{p,W}(\Omega)) := \inf_{P \in \mathcal{P}_n} \|f - P\|_{L_{p,W}(\Omega)}; \quad A_\sigma(f, L_p) := \inf_{g \in B_\sigma} \|f - g\|_{L_p(\mathbb{R})}.$$

We shall need some elementary properties of the quasi-norms and approximation errors. We set

$$\mu(p) := \min(1, p), \quad 0 < p \leq \infty.$$

Then the following inequalities hold:

(1.3.1) $\quad \|f + g\|_{L_{p,W}(E)}^{\mu(p)} \leq \|f\|_{L_{p,W}(E)}^{\mu(p)} + \|g\|_{L_{p,W}(E)}^{\mu(p)};$

(1.3.2) $\quad \|f + g\|_{L_{p,W}(E)} \leq 2^{1/\mu(p)-1}(\|f\|_{L_{p,W}(E)} + \|g\|_{L_{p,W}(E)});$

(1.3.3) $\quad E_n^{\mu(p)}(f + g, L_{p,W}(\Omega)) \leq E_n^{\mu(p)}(f, L_{p,W}(\Omega)) + E_n^{\mu(p)}(g, L_{p,W}(\Omega));$

(1.3.4) $\quad E_n(f, L_{p,W}(\Omega)) \leq \|f\|_{L_{p,W}(\Omega)}.$

Note that all these inequalities also hold if $L_{p,W}$ is replaced by the non-weighted space L_p.

1.4. Classes of Weights and Basic Estimates

Here we consider six basic classes of exponential weights $\mathcal{F}_M$, $\mathcal{F}_C$, $\mathcal{F}$, $\mathcal{F}(\text{Lip})$, $\mathcal{F}(C^2)$, and $\mathcal{F}(C^3)$ defined on a bounded or unbounded interval

(1.4.1) $\quad\quad\quad\quad I = (-c, c), \quad 0 < c \leq \infty.$

The classes $\mathcal{F}$, $\mathcal{F}(\text{Lip})$, and $\mathcal{F}(C^2)$ are defined in [72], while the definition of $\mathcal{F}(C^3)$ is given in [84].

A function $g\colon (0,c) \to (0,\infty)$ is said to be quasi-increasing if there exists $C > 0$ such that
$$g(x) \leq Cg(y), \qquad 0 < x \leq y < c.$$
Our largest clas of weights is defined below.

DEFINITION 1.4.1. Let $W = e^{-Q}$, where $Q\colon I \to [0,\infty)$ satisfies the following properties:
  (a) Q is even in I and $Q(0) = 0$.
  (b) Q' is continuous in I.
  (c) $Q' \geq 0$ in $(0,c)$.
  (d) $xQ'(x)$ is strictly increasing in $(0,c)$ and
$$\lim_{x \to c-} xQ'(x) = \infty.$$
Then we write $W \in \mathcal{F}_M$.

Next, we define a class of even convex weights, satisfying some conditions at the end points and at 0.

DEFINITION 1.4.2. Let $W = e^{-Q}$, where $Q\colon I \to [0,\infty)$ satisfies the following properties:
  (a) Q is even in I and $Q(0) = 0$.
  (b) Q' is continuous in I.
  (c) Q' is nondecreasing in I.
  (d) $\lim_{x \to c-} Q(x) = \infty$.
  (e) For any $y \in (0,c)$,
$$(1.4.2) \qquad \int_0^y \frac{Q'(s)}{s}\, ds < \infty.$$
Then we write $W \in \mathcal{F}_C$.

The next class contains weights, satisfying some growth and regularity conditions (cf. [72, p. 10]).

DEFINITION 1.4.3. Let $W = e^{-Q}$, where $Q\colon I \to [0,\infty)$ satisfies properties (a)–(d) of Definition 1.4.2, and let Q satisfy the following additional properties:
  (a) The function
$$(1.4.3) \qquad T(x) = \frac{xQ'(x)}{Q(x)}, \qquad x \neq 0$$
is quasi-increasing in $(0,c)$ with
$$(1.4.4) \qquad \Lambda := \inf_{x \in (0,c)} T(x) > 1.$$
  (b) There exists $\varepsilon_0 \in (0,1)$ such that for $y \in (0,c)$,
$$T(y) \sim T\left(y\left(1 - \frac{\varepsilon_0}{T(y)}\right)\right).$$
Then we write $W \in \mathcal{F}$.

For many of our results we need more smoothness conditions on weights. The weights from the following class satisfy the integral inequality that is close to a local Lipschitz condition of some order (cf. [72, p. 12]).

DEFINITION 1.4.4. Let $W \in \mathcal{F}$ and let Q satisfy the following property:

(a) There exist $a \in (0, \min(\Lambda-1, 1))$ and C, $\varepsilon_1 > 0$ such that for all $x \in I\setminus\{0\}$,
$$\int_{x-\frac{\varepsilon_1(x)}{T(x)}}^{x+\frac{\varepsilon_1|x|}{T(x)}} \frac{|Q'(s) - Q'(x)|}{|s-x|^{1+a}} \, ds \leq CQ'(x)\left(\frac{T(x)}{|x|}\right)^a.$$

Then we write $W \in \mathcal{F}(\text{Lip})$.

Following is the most explicit class of weights (cf. [72, p. 7]):

DEFINITION 1.4.5. Let $W = e^{-Q}$, where $Q \colon I \to [0, \infty)$ satisfies the following properties:

(a) Q is even in I and $Q(0) = 0$.
(b) Q' is continuous in I.
(c) Q'' exists and $Q''(x) > 0$ in $(0, c)$.
(d) $\lim_{x \to c-} Q(x) = \infty$.
(e) The function
$$T(x) := \frac{xQ'(x)}{Q(x)}, \qquad x \neq 0$$
is quasi-increasing in $(0, c)$ with

(1.4.5) $$\Lambda := \inf_{x \in (0,c)} T(x) > 1.$$

(f) There exists $C > 0$ such that for all $x \in (0, c)$,

(1.4.6) $$\frac{Q''(x)}{|Q'(x)|} \leq C \frac{|Q'(x)|}{Q(x)}.$$

Then we write $W \in \mathcal{F}(C^2)$.

Finally, we need the following class of Freud weights on $\mathbb{R}$ (cf. [84, p. 17]).

DEFINITION 1.4.6. Let $W = e^{-Q}$, where $Q \colon \mathbb{R} \to [0, \infty)$ satisfies the following properties:

(a) Q is an even continuous function in $\mathbb{R}$ and $Q(0) = 0$.
(b) $Q' > 0$ in $(0, \infty)$.
(c) Q'' exists in $(0, \infty)$ and Q''' exists for x large enough.
(d) The following inequalities hold:
$$C_1 \leq \frac{(xQ'(x))'}{Q'(x)} \leq C_2, \qquad x \in (0, \infty),$$
$$\frac{x^2 |Q'''(x)|}{Q'(x)} \leq C_3, \qquad x \text{ large enough.}$$

Then we write $W \in \mathcal{F}(C^3)$.

Remarks

(a) The following inclusions between these classes of weights are valid:

(1.4.7) $$\mathcal{F}_M \supseteq \mathcal{F}_C \subseteq \mathcal{F} \subseteq \mathcal{F}(\text{Lip}) \supseteq \mathcal{F}(C^2),$$

(1.4.8) $$\mathcal{F}_M \supseteq \mathcal{F}(C^3).$$

Indeed, properties (a)–(c) and (c), (d) of Definition 1.4.2 imply properties (c) and (d) of Definition 1.4.1, respectively. Thus $\mathcal{F}_M \supseteq \mathcal{F}_C$. Next, the inclusion $\mathcal{F}_C \supseteq \mathcal{F}$ follows from Proposition 1.4.7(a) given below.

To prove the inclusion $\mathcal{F}(\text{Lip}) \supseteq \mathcal{F}(C^2)$, we assume that $W \in \mathcal{F}(C^2)$. Then there exist $C_1, \varepsilon_1 > 0$ such that for any $y \in (0, c)$ and every $t \in (y - \varepsilon_1 y/T(y), y + \varepsilon_1 y/T(y))$, the following inequality holds [72, p. 68]:
$$\frac{Q'(y) - Q'(t)}{y - t} \leq C_1 \frac{Q'(y)}{y} T(y).$$
Hence choosing a number $a \in (0, \min(\Lambda - 1, 1))$, we have
$$\int_{y - \frac{\varepsilon_1 y}{T(y)}}^{y + \frac{\varepsilon_1 y}{T(y)}} \frac{|Q'(y) - Q'(t)|}{|y - t|^{1+a}} dt \leq C_1 \frac{Q'(y)}{y} T(y) \int_{y - \frac{\varepsilon_1 y}{T(y)}}^{y + \frac{\varepsilon_1 y}{T(y)}} \frac{dt}{|y - t|^a}$$
$$= \frac{2\varepsilon_1^{1-a} C_1}{1 - a} Q'(y) \left(\frac{T(y)}{y}\right)^a.$$

Thus property (a) of Definition 1.4.4 holds with $C = 2\varepsilon_1^{1-a} C_1 (1-a)^{-1}$. Moreover, $W \in \mathcal{F}$ [72, Lemma 3.3(a)]. Therefore $\mathcal{F}(\text{Lip}) \supseteq \mathcal{F}(C^2)$.

Finally, if $W \in \mathcal{F}(C^3)$, then $xQ'(x)$ is positive and increasing in $\mathbb{R}$, by properties (b) and (d). Using also Lemma 6.2 from [84], we have
$$\lim_{x \to \infty} xQ'(x) \geq \lim_{x \to \infty} Q'(1) x^{C_1} = \infty.$$
Thus $W \in \mathcal{F}_M$ and (1.4.8) follows.

(b) In the case $I = \mathbb{R}$, a weight $W \in \mathcal{F}$, satisfying the condition
$$\sup_{x \in \mathbb{R}} T(x) < \infty,$$
is called a Freud weight. A typical example of such a weight is
$$(1.4.9) \qquad W_\alpha(x) := \exp(-|x|^\alpha), \qquad \alpha > 1, \quad I = \mathbb{R}.$$
It is proved in [72, p. 64] that for every Freud weight $W(x) = e^{-Q(x)}$, Q has at most polynomial growth on $\mathbb{R}$.

The class $\mathcal{F}(C^3)$ contains Freud-type weights as well. In particular, $W_\alpha \in \mathcal{F}(C^3)$, $\alpha > 0$. We remark that $\mathcal{F}_M$ and $\mathcal{F}(C^3)$ are the only classes in (1.4.7) and (1.4.8) which contain W_α for $0 < \alpha \leq 1$ and $I = \mathbb{R}$.

(c) A weight $W \in \mathcal{F}$, satisfying the condition
$$\sup_{x \in \mathbb{R}} T(x) = \lim_{x \to \infty} T(x) = \infty,$$
is called an Erdös weight. A typical example of such a weight for $I = \mathbb{R}$ is
$$(1.4.10) \qquad W(x) = \exp\bigl(-\exp_\ell(|x|^\alpha) + \exp_\ell(0)\bigr), \qquad \alpha > 1, \ \ell \geq 1,$$
where $\exp_0(x) := x$ and
$$\exp_k(x) := \exp\bigl(\exp_{k-1}(x)\bigr), \qquad 1 \leq k \leq \ell.$$

(d) Any weight $W \in \mathcal{F}$ on a bounded interval $(-c, c)$ is an Erdös weight. Indeed, for every $\varepsilon \in (0, c)$ we have
$$\liminf_{x \to c^-} T(x) \geq \liminf_{x \to c^-} \frac{x(Q(x) - Q(c - \varepsilon))}{(x - c + \varepsilon) Q(x)} = \frac{c - \varepsilon}{\varepsilon}.$$
Letting $\varepsilon \to 0$, we get $\lim_{x \to c^-} T(x) = \infty$.

A typical example of an exponential weight $W \in \mathcal{F}$ on $I = (-1, 1)$ is

(1.4.11) $$W(x) = \exp(-\exp_\ell(1-x^2)^{-\alpha} + \exp_\ell(1)), \qquad \alpha > 0.$$

(e) More information on Examples (1.4.9)–(1.4.11) is given in Sections 10.1 and 10.5, see also [72, Section 1.6].

(f) Note that the Gegenbauer weight $W(x) = (1-x^2)^\lambda$, $\lambda > 0$, on $I = (-1, 1)$ belongs to $\mathcal{F}_C$, but $W \notin \mathcal{F}$, since the function $T(x)$ defined in (1.4.3) does not satisfy property (b) of Definition 1.4.3 in a neighborhood of ± 1.

The Mhaskar-Rakhmanov-Saff number a_n plays an important role in weighted approximation and orthogonal polynomials. Let us assume that $W = e^{-Q} \in \mathcal{F}_M$. Then $yQ'(y)$ is positive and increasing in $(0, c)$ with
$$\lim_{y \to 0+} yQ'(y) = 0, \qquad \lim_{y \to c-} yQ'(y) = \infty.$$

We define the more general number $a_t = a_t(Q) \in (0, c)$ to be the positive root of the equation

(1.4.12) $$t = \frac{2}{\pi} \int_0^1 \frac{a_t y Q'(a_t y)}{\sqrt{1-y^2}} \, dy.$$

The detailed account on properties of Q and a_t is given in [72, Chapters 2 and 3]. We summarize some of the properties in the following proposition:

PROPOSITION 1.4.7. *Let $W = e^{-Q} \in \mathcal{F}$. Then the following statements hold:*

(a) *For $s \in (0, c)$,*

(1.4.13) $$\int_0^s \frac{Q'(y)}{y} \, dy \leq C Q'(s).$$

(b) *For $s/r \geq 1$, $s, r \in (0, c)$,*

(1.4.14) $$\frac{Q'(r)}{Q'(s)} \leq \frac{T(r)}{T(s)} \left(\frac{r}{s}\right)^{\Lambda-1}.$$

(c) *a_t is a positive increasing function in $t > 0$ and for a fixed $L > 0$ and any $t > 0$,*

(1.4.15) $$Q(a_t) \sim Q(a_{Lt}) \sim t\sqrt{\frac{1}{T(a_t)}};$$

(1.4.16) $$Q'(a_t) \sim Q'(a_{Lt}) \sim \sqrt{T(a_t)};$$

(1.4.17) $$T(a_{Lt}) \sim T(a_t);$$

(1.4.18) $$a_{Lt} \sim a_t.$$

(d) *There exists $\varepsilon = \varepsilon(Q) \in (0, 1)$ such that for $t > 1$,*

(1.4.19) $$a_t \leq C t^{1-\varepsilon};$$

(1.4.20) $$T(a_t) \leq C t^{2-\varepsilon};$$

(1.4.21) $$Q(a_t) \geq C t^\varepsilon.$$

(e) *For $0 < s \leq t \leq 2s$,*

(1.4.22) $$1 \leq \frac{a_t}{a_s} \leq 1 + \frac{C}{T(a_t)}\left(\frac{t}{s} - 1\right).$$

In particular, for a fixed $b \geq 0$ and any $n \geq 1$,

$$1 \leq \frac{a_{n+b}}{a_n} \leq 1 + \frac{C}{T(a_{n+b})}\left(\frac{b}{n}\right) \leq 1 + Cb/n, \tag{1.4.23}$$

where C is independent of n and b.

(f) *For $x \in [0, a_t]$ and $t > 0$,*

$$Q'(x) \leq \frac{Ct}{a_t\sqrt{T(a_t)}}\left(\frac{1}{1 - x/a_t}\right). \tag{1.4.24}$$

(g) *For $t > 0$,*

$$\frac{2}{\pi}\int_0^1 \frac{Q'(a_t y)\sqrt{1-y^2}}{y} + \frac{t}{a_t} = \int_0^t \frac{dz}{a_z}. \tag{1.4.25}$$

(h) *For $t \to \infty$,*

$$\int_0^t \frac{dz}{a_z} \sim \frac{t}{a_t} \to \infty. \tag{1.4.26}$$

(i) *For any polynomial $P \in \mathcal{P}_n$,*

$$\|P\|_{L_{\infty},W(I)} = \|P\|_{L_{\infty},W[-a_n,a_n]}. \tag{1.4.27}$$

(j) *For any $P \in \mathcal{P}_n$,*

$$\|P\|_{L_p,W(I)} \leq C\|P\|_{L_p,W[-a_n,a_n]}, \tag{1.4.28}$$

where C is independent of n and P.

1.5. Acknowledgements

I would like to express my gratitude to D. S. Lubinsky whose provocative talk at the 8th Conference on Approximation Theory in College Station, Texas, motivated the present work. I am grateful to Yu. A. Brudnyi with whom I have spent a lot of time over the years discussing various aspects of approximation theory. In learning orthogonal polynomials and weighted approximation, I have benefitted from the discussions with S. B. Damelin, T. Kilgore, A. L. Levin, D. S. Lubinsky, K. T-R. McLaughlin, H. N. Mhaskar, E. A. Rakhmanov, and I am indebted to all of them. I am also grateful to P. Deift, A. Kroó, D. S. Lubinsky, J. Szabados, and R. S. Varga for provision of references. Also I wish to thank Kate MacDougall for her careful typing of this manuscript.

Finally, I would like to thank my wife, Natasha, for the constant heartwarming encouragement.

CHAPTER 2

Statement of Main Results

In this chapter, we give limit theorems of weighted polynomial approximation in fairly general situations (Section 2.1). These results are based on estimates for the approximation error of entire functions of exponential type (Section 2.2) and on polynomial inequalities in the complex plane (Section 2.3).

Throughout the chapter, Λ is a number, satisfying (1.4.4), and b_n is defined in (1.2.2).

2.1. Limit Theorems of Polynomial Approximation with Exponential Weights

We first discuss results on polynomial approximation in $L_{\infty,W}(I)$ of bounded functions and functions of polynomial growth on $\mathbb{R}$.

THEOREM 2.1.1. *Let $W \in \mathcal{F}(C^2)$. Then there exist a number $\delta = \delta(W) \in (0, \frac{2(\Lambda-1)}{3(\Lambda+1)})$ and a constant $C = C(W) > 0$ such that the following statements hold:*

(a) *For any $f \in L_\infty(\mathbb{R})$ and every $\sigma > 0$,*

$$(2.1.1) \qquad \lim_{n \to \infty} E_n\bigl(f(\gamma_n \cdot), L_{\infty,W}(I)\bigr) = A_\sigma(f, L_\infty);$$

$$(2.1.2) \qquad \lim_{n \to \infty} E_n\bigl(f(\gamma_n \cdot), L_{\infty,W}[-a_n, a_n]\bigr) = A_\sigma(f, L_\infty);$$

where

$$(2.1.3) \qquad \gamma_n := \frac{b_n(1 - Cn^{-\delta})}{\sigma}.$$

(b) *Let a continuous function f satisfy the conditions*

$$(2.1.4) \qquad |f(x)| \leq A(1 + |x|)^N, \qquad x \in \mathbb{R}, \quad N \geq 0.$$

and

$$(2.1.5) \qquad A_{\sigma_0}(f, L_\infty) < \infty$$

for some $\sigma_0 > 0$. Then relations (2.1.1) and (2.1.2) hold for all $\sigma \geq \sigma_0$.

The next result shows that for $0 < p < \infty$, an L_p-analogue of (2.1.1) and (2.1.2) is valid with γ_n replaced by b_n/σ.

THEOREM 2.1.2. *For $W \in \mathcal{F}(C^2)$ the following statements hold:*

(a) *For any $f \in L_p(\mathbb{R})$, $0 < p < \infty$, and every $\sigma > 0$,*

$$(2.1.6) \qquad \lim_{n \to \infty} (b_n/\sigma)^{1/p} E_n\bigl(f((b_n/\sigma)\cdot), L_{p,w}(I)\bigr) = A_\sigma(f, L_p);$$

$$(2.1.7) \qquad \lim_{n \to \infty} (b_n/\sigma)^{1/p} E_n\bigl(f((b_n/\sigma)\cdot), L_{p,W}[-a_n, a_n]\bigr) = A_\sigma(f, L_p).$$

(b) *Let a measurable f satisfy condition (2.1.4) and let*

(2.1.8) $$A_{\sigma_0}(f, L_p) < \infty, \qquad 0 < p < \infty,$$

for some $\sigma_0 > 0$. Then relations (2.1.6) and (2.1.7) hold for all $\sigma > \sigma_0$.

Next, we consider certain analogues of Theorems 2.1.1 and 2.1.2 for more general classes of functions.

THEOREM 2.1.3. *Let $W \in \mathcal{F}(C^2)$. Then there exists a number $\beta = \beta(W) \in [1, \infty)$ such that the following statements hold:*
  (a) *Let $\delta(W)$ and $C(W)$ be the constants from Theorem 2.1.1. Then for any $f \in L_{\infty,\text{loc}}(\mathbb{R})$, satisfying condition (2.1.5) for some $\sigma_0 > 0$, and for every $\sigma > \beta\sigma_0$, relation (2.1.2) is valid.*
  (b) *For any $f \in L_{p,\text{loc}}(\mathbb{R})$, $0 < p < \infty$, satisfying (2.1.8) for some $\sigma_0 > 0$, and for every $\sigma > \beta\sigma_0$, relation (2.1.7) is valid.*

For Erdös weights, it is possible to give an upper bound for β.

THEOREM 2.1.4. *Let $W \in \mathcal{F}(C^2)$ with $\lim_{x \to \infty} T(x) = \infty$. Then statements (a) and (b) of Theorem 2.1.3 hold with $\beta \in [1, 3]$.*

Remarks

(a) Limit theorems of polynomial approximation for some special weights are established in Chapter 10. Various applications of these results are discussed in Chapter 8. The proofs of Theorems 2.1.1–2.1.4 are given in Chapter 7.

(b) The constant b_n defined in (1.2.2) plays an important role in the limit theorems as well as in polynomial inequalities and some applications. We show in Chapter 3 that
$$b_n \sim n/a_n$$
and for Erdös weights,
$$b_n = (n/a_n)(1 + o(1)), \qquad n \to \infty.$$

(c) Some sufficient conditions on f for (2.1.5) and (2.1.8) to hold are given in Lemma 8.1.5.

(d) Theorem 2.1.2(b) for $I = \mathbb{R}$, $W(x) = x^2$, and $p \in [1, 2]$ was proved in [50].

2.2. Approximation of Entire Functions of Exponential Type

In the early 1980s, Mhaskar [89, 90] proved the following estimate:

(2.2.1) $$\limsup_{n \to \infty} \left(\frac{n! E_n(g, L_{p,W}(\mathbb{R}))}{s_n} \right)^{1/n} \leq C\sigma, \qquad 1 \leq p \leq \infty,$$

where $g \in B_\sigma$ and $W = e^{-Q}$ satisfies some restrictive Freud-type conditions. In particular, W_α satisfies these conditions for $\alpha \geq 2$. Here, $s_n = \prod_{p=1}^{n} q_p$ and q_p is the least positive solution to the equation $xQ'(x) = p$. Inequality (2.2.1) was extended in [91, p. 187] to all Freud weights from $\mathcal{F}(C^2)$ for $I = \mathbb{R}$.

To prove the limit theorems, we need estimates of the approximation error in $L_{p,W}(I)$, $0 < p \leq \infty$, of an entire function of varying exponential type $\sigma \leq b_n(1 - \varepsilon_n)$ with more general weights on bounded and unbounded intervals.

We first consider entire functions of exponential type that are of at most polynomial growth on $\mathbb{R}$.

THEOREM 2.2.1. *Let $W \in \mathcal{F}(C^2)$. Then there exist numbers $\delta_1 = \delta_1(W) \in \left(0, \frac{2(\Lambda-1)}{3(\Lambda+1)}\right)$, $\delta_2 = \delta_2(W) > 0$, and a constant $C_1 = C_1(W) > 0$ such that for every $\sigma \in \left(0, b_n(1 - C_1 n^{-\delta_1})\right]$ and any $g \in B_\sigma$, satisfying the inequality*

$$(2.2.2) \qquad |g(x)| \leq A n^m (1 + |x|)^N, \qquad x \in \mathbb{R},$$

the following estimate holds:

$$(2.2.3) \qquad E_n\bigl(g, L_{p,W}(I)\bigr) \leq C_2 A n^\gamma \exp\bigl(-C_3 n^{\delta_2}\bigr), \qquad 0 < p \leq \infty.$$

Here, A, m, N are nonnegative constants independent of x and n, and C_2, C_3, γ are positive constants independent of n and g.

Next, we consider an analogue of this theorem for any entire function of exponential type.

THEOREM 2.2.2. (a) *Let $W \in \mathcal{F}(C^2)$. Then there exists a number $\beta = \beta(W) \in [1, \infty)$ such that for all $\varepsilon > 0$ small enough, every $\sigma \in \bigl(0, b_n/(\beta(1+\varepsilon))\bigr)$, and for any $g \in B_\sigma$, the inequality*

$$(2.2.4) \qquad E_n\bigl(g, L_{p,W}[-a_n, a_n]\bigr) \leq C n^\gamma \exp(-C_1 \varepsilon n), \qquad 0 < p \leq \infty,$$

holds. Here $C = C(g, \varepsilon)$, and C_1, $\gamma \geq 0$ are independent of n and g.

(b) *Let $W \in \mathcal{F}(C^2)$ with $\lim_{x \to \infty} T(x) = \infty$. Then the constant β in statement* (a) *satisfies the inequalities $1 < \beta < 3$.*

Remarks

(a) It is easy to verify that if f is an entire function of exponential type $\sigma \geq \sigma_0$, satisfying (2.1.4), then Theorems 2.1.1 and 2.1.2 follow from Theorem 2.2.1. Also, Theorem 2.2.2 implies Theorems 2.1.3 and 2.1.4 for any $f \in B_\sigma$, $\sigma > \beta \sigma_0$.

(b) The analytic expression for $\beta(W)$ in Theorems 2.1.3, 2.1.4, and 2.2.2 is given by (6.4.13). We conjecture that this is the least constant such that these theorems hold. It is possible to compute $\beta(W)$ explicitly for $W_\alpha(x) = \exp(-|x|^\alpha)$, $\alpha > 1$ (see Theorem 10.1.2). The corresponding constant $\gamma = 1.5088\ldots$ for non-weighted approximation was found by Bernstein [16]. Note that $\lim_{\alpha \to \infty} \beta(W_\alpha) = \gamma$ (see Remark (c) in Section 10.1).

(c) A weaker version of Theorem 2.2.1 for $I = \mathbb{R}$, $W(x) = \exp(-x^2)$, and $p \in [1, \infty]$ was proved by a different method in [50].

(d) The proofs of Theorems 2.2.1 and 2.2.2 are given in Chapter 6. We remark that these theorems do not follow from (2.2.1) even for the weights considered in [89–91].

2.3. Polynomial Inequalities in the Complex Plane

A V. A. Markov-type inequality for the coefficients of a polynomial $P_n(x) = \sum_{k=0}^{n} c_k x^k$,

$$(2.3.1) \qquad |c_k| \leq \frac{n^k}{k!} \|P_n\|_{L_\infty[-1,1]}, \qquad 0 \leq k \leq n,$$

(see [114, p. 227]) plays an essential role in the proof of non-weighted limit theorem (1.1.4). It easily follows from (2.3.1) that

$$|P_n(z)| \leq e^{n|z|} \|P_n\|_{L_\infty[-1,1]}, \qquad z \in \mathbb{C}. \tag{2.3.2}$$

On the other hand, this estimate can be used to prove a version of (2.3.1). Indeed, (2.3.2) follows from the Bernstein inequality [114, p. 91]

$$\begin{aligned} |P_n(z)| &\leq |\psi(z)|^n \|P_n\|_{L_\infty[-1,1]} \leq \left(|z| + \sqrt{|z|^2 + 1}\right)^n \|P_n\|_{L_\infty[-1,1]} \\ &\leq e^{n|z|} \|P_n\|_{L_\infty[-1,1]}, \qquad z \in \mathbb{C}. \end{aligned}$$

Then using Cauchy's integral theorem, we arrive at an inequality like (2.3.1).

We use the similar approach to prove V. A. Markov-type inequalities for weighted metrics. We first establish a weighted version of inequality (2.3.2).

THEOREM 2.3.1. *Let $W \in \mathcal{F}_C$. Then for any polynomial $P \in \mathcal{P}_n$,*

$$|P(z)| \leq e^{b_n|z|} \|P\|_{L_{\infty,W}(I)}, \qquad z \in \mathbb{C}. \tag{2.3.3}$$

The proof is based on the Bernstein-type inequality for exponential weights and on inequality (1.2.4) whose proof is quite technical.

As a corollary of Theorem 2.3.1, we obtain a weighted version of V. A. Markov's inequality.

THEOREM 2.3.2. *Let $W \in \mathcal{F}_C$. Then for any polynomial $P(x) = \sum_{k=0}^{n} c_k x^k$ the following statements hold:*

(a)

$$|c_k| \leq \frac{3\sqrt{k+1}\, b_n^k}{k!} \|P\|_{L_{\infty,W}[-a_n, a_n]}, \qquad 0 \leq k \leq n. \tag{2.3.4}$$

(b) *For every $\varepsilon \in (0,1)$ and $p \in (0, \infty)$,*

$$|c_k| \leq \frac{C\sqrt{k+1}\, b_n^{k+1/p}}{(1-\varepsilon)^k k!} \|P\|_{L_{p,W}[-a_n, a_n]}, \qquad 0 \leq k \leq n, \tag{2.3.5}$$

where C is independent of k, n, and P.

The next result shows that the constant b_n in (2.3.3) cannot be replaced by $(1-\mu)b_n$, $\mu > 0$, for all polynomials $P \in \mathcal{P}_n$, $n = 1, 2, \ldots$, and all $z \in \mathbb{C}$.

THEOREM 2.3.3. *If $W \in \mathcal{F}(C^2)$, then*

$$\lim_{n \to \infty} b_n^{-1} \sup_{P \in \mathcal{P}_n} \sup_{z \in \mathbb{C}} |z|^{-1} \log \frac{|P(z)|}{\|P\|_{L_{\infty,W}(I)}} = 1. \tag{2.3.6}$$

Remarks

(a) Theorem 2.3.2(b) for $I = \mathbb{R}$, $W(x) = \exp(-x^2)$, and $p \in (0, 2]$ was proved by a different method in [50].

(b) The proofs of Theorems 2.3.1 and 2.3.2 are given in Section 4.4, while Theorem 2.3.3 is proved in Section 10.1.

CHAPTER 3

Properties of Harmonic Functions

In this chapter, we establish some properties of the integral

$$(3.0.7) \qquad H(w) = H(W,w) := \frac{1}{2\pi}\int_{-\pi}^{\pi} Q(a_n \cos t)\frac{1+we^{it}}{1-we^{et}}\,dt, \qquad |w|<1,$$

where $Q(a_n \cos t) = -\log W(a_n \cos t)$. We also introduce certain functions and constants, related to this integral, and study their properties.

We remark that $\exp(H(W,w))$ coincides with $D^{-2}(W,w)$, where $D(W,w)$ is the Szegö function (cf. [72, p. 2]).

3.1. The Poisson Integral Re $H(w)$

The following proposition contains some elementary properties of harmonic functions.

PROPOSITION 3.1.1. *Let* $W = e^{-Q} \in \mathcal{F}_M$. *Then the following statements hold:*

(a) *For* $w = re^{i\varphi}$, $0 \le r < 1$, $\varphi \in [0, 2\pi)$, *we have*

$$(3.1.1) \qquad \operatorname{Re} H(w) = \frac{1-r^2}{2\pi}\int_{-\pi}^{\pi} \frac{Q(a_n \cos t)}{1+r^2-2r\cos(t+\varphi)}\,dt.$$

(b) $\operatorname{Re} H(w)$ *is a harmonic function for* $|w| < 1$.

(c) *For any* $\varphi \in [0, 2\pi)$,

$$\lim_{w \to e^{i\varphi}, |w|<1} \operatorname{Re} H(w) = Q(a_n \cos \varphi);$$

in particular,

$$(3.1.2) \qquad H(e^{i\varphi}) := \lim_{r \to 1-} \operatorname{Re} H(re^{i\varphi}) = Q(a_n \cos \varphi).$$

(d) *For each* $r \in [0,1]$, $\operatorname{Re} H(re^{i\varphi})$ *is an even and π-periodic function in* $\varphi \in [-\pi, \pi]$.

PROOF. Statements (a)–(c) are the well-known facts about harmonic functions [109]. To prove statement (d) for $r \in [0, 1)$, we first note that $\operatorname{Re} H(re^{-i\varphi}) = \operatorname{Re} H(re^{i\varphi})$, by (3.1.1). Then taking account of property (a) of Definition 1.4.1, we have from (3.1.1)

$$\operatorname{Re} H\bigl(re^{i(\varphi+\pi)}\bigr) = \frac{1-r^2}{2\pi}\int_{-\pi}^{\pi}\frac{Q(-a_n \cos(t-\varphi))}{1+r^2-2r\cos t}\,dt = \operatorname{Re} H\bigl(re^{i\varphi}\bigr).$$

Thus (d) follows for $r \in [0,1)$, while for $r = 1$ (3.1.2) implies (d). ∎

Next, we need some properties of $H(ir)$ and $H'(ir)$.

PROPOSITION 3.1.2. *Let $W \in \mathcal{F}_M$ unless indicated otherwise.*

(a) $H(ir)$ *is a real-valued function for* $0 \leq r \leq 1$.

(b) *For* $0 \leq r < 1$,

$$H(ir) = \frac{1-r^2}{2\pi} \int_{-\pi}^{\pi} \frac{Q(a_n \cos t)}{1+r^2+2r\sin t} \, dt \tag{3.1.3}$$

and

$$H(i) = 0. \tag{3.1.4}$$

(c) *For* $0 \leq r < 1$,

$$\begin{aligned} iH'(ir) &= \frac{a_n}{\pi} \int_{-\pi}^{\pi} \frac{Q'(a_n \sin t)\sin t \cos t}{1+r^2+2r\cos t} \, dt \\ &= \frac{a_n}{\pi} \int_{-\pi}^{\pi} \frac{Q'(a_n \cos t)\sin t \cos t}{1+r^2+2r\sin t} \, dt. \end{aligned} \tag{3.1.5}$$

(d) *Let* $W \in \mathcal{F}_C$. *Then we have*

$$\begin{aligned} iH'(i) &:= i \lim_{r \to 1-} H'(ir) = \frac{a_n}{2\pi} \int_{-\pi}^{\pi} \frac{Q'(a_n \sin t)\sin t \cos t}{1+\cos t} \, dt \\ &= \frac{a_n}{2\pi} \int_{-\pi}^{\pi} \frac{Q'(a_n \cos t)\sin t \cos t}{1+\sin t} \, dt \\ &= -\frac{2a_n}{\pi} \int_0^1 \frac{Q'(a_n x)\sqrt{1-x^2}}{x} \, dx. \end{aligned} \tag{3.1.6}$$

PROOF. (a) We have from (3.0.7) for $0 \leq r < 1$

$$\begin{aligned} \operatorname{Im} H(ir) &= \frac{r}{2\pi} \int_{-\pi}^{\pi} \frac{Q(a_n \cos t)\cos t}{1+r^2+2r\sin t} \, dt \\ &= \frac{r}{2\pi} \int_{-\pi}^{\pi} \frac{Q(a_n \sin t)\sin t}{1+r^2-2r\cos t} \, dt = 0, \end{aligned}$$

since $Q(a_n \sin t) \sin t$ is an odd function, and $\operatorname{Im} H(i) = 0$, by (3.1.2).

(b) Integral representation (3.1.3) immediately follows from (3.1.1) and statement (a), while (3.1.4) is a consequence of (3.1.2) and the property $Q(0) = 0$.

(c) Differentiating the formula

$$H(ir) = \frac{1}{2\pi} \int_{-\pi}^{\pi} Q(a_n \cos t) \left(\frac{2}{1-ire^{it}} - 1\right) dt, \qquad 0 \leq r < 1,$$

and then integrating by parts, we get

$$\begin{aligned} \frac{dH(ir)}{dr} &= \frac{i}{\pi} \int_{-\pi}^{\pi} \frac{Q(a_n \cos t)e^{it}}{(1-ire^{it})^2} \, dt = -\frac{ia_n}{\pi r} \int_{-\pi}^{\pi} \frac{Q'(a_n \cos t)\sin t}{1-ire^{it}} \, dt \\ &= -\frac{ia_n}{\pi r} \int_{-\pi}^{\pi} \frac{Q'(a_n \cos t)\sin t(1+r\sin t+ir\cos t)}{1+r^2+2r\sin t} \, dt \\ &= -\frac{ia_n}{\pi r} \int_{-\pi}^{\pi} \frac{Q'(a_n \sin t)\cos t(1+r\cos t+ir\sin t)}{1+r^2+2r\cos t} \, dt. \end{aligned}$$

This implies (3.1.5) since $Q'(a_n \sin t)$ is an odd function.

(d) Let $\{r_k\}_{k=1}^\infty$ be a sequence of numbers from $[1/2, 1)$ such that $\lim_{k\to\infty} r_k = 1$. It is easy to see that for a sequence of functions

$$f_k(t) := \frac{Q'(a_n \sin t) \sin t \cos t}{1 + r_k^2 + 2r_k \cos t}, \qquad k = 1, 2, \ldots,$$

we have

(3.1.7) $$\lim_{k\to\infty} f_k(t) = f_0(t) := \frac{Q'(a_n \sin t) \sin t \cos t}{2(1 + \cos t)}, \qquad t \in (-\pi, \pi).$$

Next, for all $t \in (-\pi, \pi)$,

(3.1.8) $$|f_k(t)| \leq \frac{|Q'(a_n \sin t) \sin t \cos t|}{2r_k(1 + \cos t)} \leq 2|f_0(t)|, \qquad k = 1, 2, \ldots.$$

Further, $|f_0|$ is integrable on $[-\pi, \pi]$ since by (1.4.2),

$$\int_{-\pi}^\pi |f_0(t)|\, dt \leq \int_{-\pi}^\pi \frac{|Q'(a_n \sin t) \sin t|}{2(1 - |\cos t|)}\, dt \leq 2 \int_0^\pi \frac{Q'(a_n \sin t)}{\sin t}\, dt$$

$$= 4 \int_0^1 \frac{Q'(a_n y)}{y\sqrt{1-y^2}}\, dy = 4\left(\int_0^{1/2} + \int_{1/2}^1\right)$$

$$\leq C \int_0^{1/2} \frac{Q'(a_n y)}{y}\, dy + C_1 Q'(a_n) < \infty.$$

Then by the Lebesgue dominated convergence theorem, we get from (3.1.5), (3.1.7), and (3.1.8)

(3.1.9) $$i \lim_{r\to 1-} H'(ir) = \frac{a_n}{2\pi} \int_{-\pi}^\pi \frac{Q'(a_n \sin t) \sin t \cos t}{1 + \cos t}\, dt.$$

Finally using the facts that $Q'(x)$ is an odd function in x and $Q'(a_n \cos t)$ an even function in t, we have from (3.1.9)

$$\begin{aligned}
iH'(i) &= \frac{a_n}{2\pi} \int_{-\pi}^\pi \frac{Q'(a_n \cos t) \sin t \cos t}{1 + \sin t}\, dt \\
&= \frac{a_n}{2\pi} \int_{-\pi}^\pi Q'(a_n \cos t) \cos t \left(\frac{\sin t}{1 - \sin^2 t} - \frac{\sin^2 t}{1 - \sin^2 t}\right) dt \\
&= -\frac{a_n}{\pi} \int_0^\pi \frac{Q'(a_n \cos t) \sin^2 t}{\cos t}\, dt = -\frac{2a_n}{\pi} \int_0^{\pi/2} \frac{Q'(a_n \cos t) \sin^2 t}{\cos t}\, dt \\
&= -\frac{2a_n}{\pi} \int_0^1 \frac{Q'(a_n x)\sqrt{1-x^2}}{x}\, dx.
\end{aligned}$$

This completes the proof of (3.1.6). ∎

PROPOSITION 3.1.3. *If $W \in \mathcal{F}_M$, then for $w \in \mathbb{C}$ and $0 \leq r < 1$,*

(3.1.10) $$\min_{|w|=r} \operatorname{Re} H(w) = H(ir).$$

PROOF. Let $w = re^{i\varphi}$, $0 < r < 1$, $\varphi \in [-\pi, \pi]$, and let

$$L(\varphi) := \frac{2\pi}{a_n(1-r^2)} \operatorname{Re} H(re^{i\varphi}) = \frac{1}{a_n} \int_{-\pi}^\pi \frac{Q(a_n \cos(t-\varphi))}{1 + r^2 - 2r\cos t}\, dt.$$

Then we obtain by a straightforward calculation,

$$\begin{aligned}
\frac{dL(\varphi)}{d\varphi} &= \int_{-\pi}^{\pi} \frac{Q'(a_n \cos t) \sin t}{1+r^2-2r\cos(t+\varphi)} \, dt = \int_0^{\pi/2} Q'(a_n \cos t) \sin t \\
&\quad \times \left(\frac{1}{1+r^2-2r\cos(t+\varphi)} - \frac{1}{1+r^2-2r\cos(t-\varphi)} \right. \\
&\quad \left. - \frac{1}{1+r^2-2r\cos(t-\varphi)} + \frac{1}{1+r^2+2r\cos(t+\varphi)} \right) dt \\
&= 8r^2(1+r^2) \int_0^{\pi/2} Q'(a_n \cos t) \sin t \\
&\quad \times \frac{\cos^2(t+\varphi) - \cos^2(t-\varphi)}{\bigl((1+r^2)^2 - (2r\cos(t+\varphi))^2\bigr)\bigl((1+r^2)^2 - (2r\cos(t-\varphi))^2\bigr)} \, dt \\
&= -8r^2(1+r^2) \int_0^{\pi/2} Q'(a_n \cos t) \\
&\quad \times \frac{\sin t \sin 2t \sin 2\varphi}{\bigl(1+r^4-2r^2\cos(2(t+\varphi))\bigr)\bigl(1+r^4-2r^2\cos(2(t-\varphi))\bigr)} \, dt.
\end{aligned}$$

Hence for $0 < r < 1$

$$L'(\varphi) = -\sin 2\varphi \, M(\varphi),$$

where by property (c) of Definition 1.4.1, $M(\varphi) > 0$ for all $\varphi \in [-\pi, \pi]$. Moreover, by Proposition 3.1.1(d), $L(\varphi)$ and $L'(\varphi)$ are π-periodic functions. Thus $L(\varphi)$ attains its minimum at $\varphi = \pi/2$, and (3.1.10) follows from Proposition 3.1.2(a). ∎

The following is a new integral representation for $\operatorname{Re} H(w)$.

PROPOSITION 3.1.4. *Let $W \in \mathcal{F}_M$. Then for $0 \le \varphi \le \pi/2$ and $0 \le r < 1$ we have*

$$\operatorname{Re} H(re^{i\varphi}) = -\frac{a_n}{\pi} \int_0^{\pi/2} Q'(a_n \cos t) \sin t \, K(r,t,\varphi) \, dt, \tag{3.1.11}$$

where
(3.1.12)
$$K(r,t,\varphi) :=$$
$$\begin{cases} \arctan\left(\frac{1-r^2}{1+r^2} \cot(t+\varphi)\right) + \arctan\left(\frac{1-r^2}{1+r^2} \cot(t-\varphi)\right), & 0 \le t < \varphi \le \frac{\pi}{2} \\ \arctan\left(\frac{1-r^2}{1+r^2} \cot(t+\varphi)\right) + \arctan\left(\frac{1-r^2}{1+r^2} \cot(t-\varphi)\right) - \pi, & \frac{\pi}{2} \ge t > \varphi \ge 0 \\ \arctan\left(\frac{1-r^2}{1+r^2} \cot(2\varphi)\right) - \frac{\pi}{2}, & t = \varphi. \end{cases}$$

PROOF. Note first that integration by parts in the right-hand side of (3.1.1) gives

$$\begin{aligned}
\operatorname{Re} H(re^{i\varphi}) &= \frac{1}{\pi} Q(a_n \cos t) K_1(r,t,\varphi) \Big|_{t=-\pi}^{\pi} \\
&\quad + \frac{a_n}{\pi} \int_{-\pi}^{\pi} Q'(a_n \cos t) \sin t \, K_1(r,t,\varphi) \, dt,
\end{aligned} \tag{3.1.13}$$

where $\varphi \in [0, \pi/2]$ and

(3.1.14) $\quad K_1(r,t,\varphi) := \begin{cases} \arctan\left(\frac{1+r}{1-r} \tan \frac{t+\varphi}{2}\right), & -\pi \leq t < \pi - \varphi \\ \arctan\left(\frac{1+r}{1-r} \tan \frac{t+\varphi}{2}\right) + \pi, & \pi - \varphi < t \leq \pi \\ \pi/2 & t = \pi - \varphi. \end{cases}$

Next we have from (3.1.14)

(3.1.15) $\quad K_1(r, \pi, \varphi) - K_1(r, -\pi, \varphi) = \pi.$

Then (3.1.13) and (3.1.15) imply the representation

(3.1.16) $\quad \operatorname{Re} H(re^{i\varphi}) = Q(a_n) + \frac{a_n}{\pi} \int_{-\pi}^{\pi} Q'(a_n \cos t) \sin t\, K_1(r,t,\varphi)\, dt.$

To transform the interval of integration $[-\pi, \pi]$ into $[0, \pi/2]$, we shall use the formula

(3.1.17) $\quad \int_{-\pi}^{\pi} Q'(a_n \cos t) F(t)\, dt = \int_{0}^{\pi/2} Q'(a_n \cos t) A(F,t)\, dt,$

where the linear operator A is given by

(3.1.18) $\quad A(F,t) := F(t) + F(-t) - F(\pi - t) - F(-\pi + t), \quad t \in [0, \pi/2].$

Further for $\varphi \in [0, \pi/2]$ we obtain from (3.1.14) and (3.1.18)

$A\big(\sin t\, K_1(r,t,\varphi), t\big)$
$= \sin t \big(K_1(r,t,\varphi) - K_1(r,-t,\varphi) - K_1(r,\pi-t,\varphi) + K_1(r,-\pi+t,\varphi)\big)$
$= \sin t \Bigg(\arctan\left(\frac{1+r}{1-r} \tan \frac{t+\varphi}{2}\right) + \arctan\left(\frac{1+r}{1-r} \tan \frac{t-\varphi}{2}\right)$

$- \begin{cases} \arctan\left(\frac{1+r}{1-r} \cot \frac{t-\varphi}{2}\right), & \varphi < t \\ \arctan\left(\frac{1+r}{1-r} \cot \frac{t-\varphi}{2}\right) + \pi, & \varphi > t \\ \pi/2, & \varphi = t \end{cases}$

(3.1.19) $\quad - \arctan\left(\frac{1+r}{1-r} \cot \frac{t+\varphi}{2}\right) \Bigg).$

Using the elementary identity

$$\arctan a - \arctan b = \arctan \frac{a-b}{1+ab}$$

for $ab > 1$ with

$$a = \frac{1+r}{1-r} \tan \frac{t \pm \varphi}{2}, \quad b = \frac{1+r}{1-r} \cot \frac{t \pm \varphi}{2}, \quad ab = \left(\frac{1+r}{1-r}\right)^2,$$

we have by simple calculations,

$\sin t \left(\arctan\left(\frac{1+r}{1-r} \tan \frac{t \pm \varphi}{2}\right) - \arctan\left(\frac{1+r}{1-r} \cot \frac{t \pm \varphi}{2}\right) \right)$

(3.1.20) $\quad = -\sin t \arctan\left(\frac{1-r^2}{1+r^2} \cot \frac{t \pm \varphi}{2}\right).$

Then combining the first term in the right-hand side of (3.1.19) with the fourth one and the second term with the third one, we have from (3.1.19) and (3.1.20)

$$A\bigl(\sin t\, K_1(r,t,\varphi), t\bigr) = -\sin t\bigl(K(r,t,\varphi) + \pi\bigr). \tag{3.1.21}$$

Finally, using (3.1.16)–(3.1.18) and (3.1.21), we get

$$\begin{aligned}
\operatorname{Re} H(re^{i\varphi}) &= \frac{a_n}{\pi} \int_0^{\pi/2} Q'(a_n \cos t)(\pi \sin t)\, dt \\
&\quad - \frac{a_n}{\pi} \int_0^{\pi/2} Q'(a_n \cos t) \sin t \bigl(K(r,t,\varphi) + \pi\bigr) dt \\
&= \frac{a_n}{\pi} \int_0^{\pi/2} Q'(a_n \cos t) \sin t\, K(r,t,\varphi)\, dt.
\end{aligned}$$

Thus (3.1.11) follows. ■

In particular, the following statement immediately follows from Proposition 3.1.4:

PROPOSITION 3.1.5. *If $W \in \mathcal{F}_M$, then for $0 \le r < 1$,*

$$H(ir) = \frac{2a_n}{\pi} \int_0^{\pi/2} Q'(a_n \cos t) \sin t \arctan\left(\frac{1-r^2}{1+r^2} \tan t\right) dt. \tag{3.1.22}$$

3.2. The Function $h(r)$ and the Constant b_n

Basic properties of the function

$$h(r) = h(Q, n, r) := -\frac{2r\bigl(ir H'(ir) - n\bigr)}{a_n(1+r^2)} \tag{3.2.1}$$

are given in the following proposition:

PROPOSITION 3.2.1. *Let $W \in \mathcal{F}_M$ unless indicated otherwise.*

(a) *For $0 \le r \le 1$,*

$$h(r) = \frac{2r}{\pi} \int_0^\pi \frac{Q'(a_n \sin t) \sin t}{1 + r^2 + 2r \cos t}\, dt. \tag{3.2.2}$$

(b) *For $0 \le r < 1$,*

$$h'(r) = \frac{2(1-r^2)}{\pi} \int_0^\pi \frac{Q'(a_n \sin t) \sin t}{(1+r^2+2r\cos t)^2}\, dt > 0. \tag{3.2.3}$$

(c) *h is a positive, increasing and continuous function in $r \in [0,1)$ and $h(0) = 0$.*

(d) *If $W \in \mathcal{F}_C$, then*

$$h(1) := \lim_{r \to 1-} h(r) = b_n = \frac{1}{\pi} \int_0^\pi \frac{Q'(a_n \sin t) \sin t}{1 + \cos t}\, dt$$

$$= \frac{2}{\pi} \int_0^{\pi/2} \frac{Q'(a_n \sin t)}{\sin t}\, dt = \frac{2}{\pi} \int_0^{\pi/2} \frac{Q'(a_n \cos t)}{\cos t}\, dt, \tag{3.2.4}$$

where b_n is defined in (1.2.2).

3.2. THE FUNCTION $h(r)$ AND THE CONSTANT b_n

PROOF. Using (1.4.12), (3.2.1), (3.1.5), and (3.1.6), we have

$$h(r) = -\frac{2r}{a_n(1+r^2)}\left(\frac{a_n r}{\pi}\int_{-\pi}^{\pi}\frac{Q'(a_n\cos t)\sin t\cos t}{1+r^2+2r\sin t}dt\right.$$
$$\left. - \frac{a_n}{2\pi}\int_{-\pi}^{\pi}Q'(a_n\cos t)\cos t\,dt\right)$$
$$= \frac{r}{\pi}\int_{-\pi}^{\pi}\frac{Q'(a_n\cos t)\cos t}{1+r^2+2r\sin t}dt = \frac{r}{\pi}\int_{-\pi}^{\pi}\frac{Q'(a_n\sin t)\sin t}{1+r^2+2r\cos t}dt.$$

This yields (3.2.2). Next differentiating (3.2.2) with respect to r, we obtain for $0 \leq r < 1$

$$h'(r) = \frac{2}{\pi}\int_0^{\pi}Q'(a_n\sin t)\left(\frac{1}{1+r^2+2r\cos t} - \frac{r(2r+2\cos t)}{(1+r^2+2r\cos t)^2}\right)\sin t\,dt$$
$$= \frac{2(1-r^2)}{\pi}\int_0^{\pi}\frac{Q'(a_n\sin t)\sin t}{(1+r^2+2r\cos t)^2}dt.$$

Hence (3.2.3) holds. Next, statement (c) immediately follows from (3.2.2) and (3.2.3). Finally, using (3.2.2) for $r = 1$ and (1.4.12), we have

$$h(1) = \frac{1}{\pi}\int_0^{\pi}\frac{Q'(a_n\sin t)\sin t}{1+\cos t}dt$$
$$= \frac{1}{\pi}\int_0^{\pi/2}Q'(a_n\sin t)\left(\frac{1}{1+\cos t}+\frac{1}{1-\cos t}\right)\sin t\,dt$$
$$= \frac{2}{\pi}\int_0^{\pi/2}\frac{Q'(a_n\sin t)}{\sin t}dt = \frac{2}{\pi}\int_0^1\frac{Q'(a_n y)}{y\sqrt{1-y^2}}dy = b_n.$$

Thus (3.2.4) follows. ∎

In the following five propositions, we obtain some technical estimates for b_n and h.

PROPOSITION 3.2.2. *Let $W \in \mathcal{F}$. Then the following statements hold:*

(a)

(3.2.5) $$\frac{n}{a_n} < b_n \leq \frac{n}{a_n}\left(1+\frac{C}{\sqrt{T(a_n)}}\right) \leq (1+C_1)\frac{n}{a_n}.$$

(b) *If $T(a_n) \sim C$ (Freud weights, $I = \mathbb{R}$), then*

(3.2.6) $$\frac{n}{a_n} < b_n \leq (1+C_1)\frac{n}{a_n}.$$

(c) *If $\lim_{n\to\infty} T(a_n) = \infty$ (Erdös weights), then*

(3.2.7) $$b_n = \frac{n}{a_n}(1+o(1)), \quad n \to \infty.$$

(d)

(3.2.8) $$\lim_{n\to\infty} n/a_n = \lim_{n\to\infty} b_n = \infty.$$

PROOF. It is easy to verify that statements (b) and (c) immediately follow from statement (a), while (1.4.25), (1.4.26), and (3.2.5) imply (3.2.8). Besides, the left inequality in (3.2.5) is trivial and the right one follows from the property $T(a_n) \geq \Lambda > 1$. Consequently it remains to prove the estimate

$$\text{(3.2.9)} \qquad \int_0^1 \frac{Q'(a_n y)\sqrt{1-y^2}}{y}\,dy \leq \frac{Cn}{a_n\sqrt{T(a_n)}}.$$

We first note that

$$\text{(3.2.10)} \qquad \int_0^1 \frac{Q'(a_n y)\sqrt{1-y^2}}{y}\,dy \leq \int_0^1 \frac{Q'(a_n y)}{y}\,dy = \int_0^{1/2} + \int_{1/2}^1.$$

Next, we use (1.4.13) and (1.4.24) to show that

$$\text{(3.2.11)} \qquad \int_0^{1/2} \frac{Q'(a_n y)}{y}\,dy \leq CQ'(a_n/2) \leq \frac{Cn}{a_n\sqrt{T(a_n)}}.$$

Further by (1.4.15),

$$\text{(3.2.12)} \qquad \int_{1/2}^1 \frac{Q'(a_n y)}{y}\,dy \leq \frac{2}{a_n}(Q(a_n) - Q(a_n/2)) \leq \frac{Cn}{a_n\sqrt{T(a_n)}}.$$

Thus (3.2.10)–(3.2.12) yield (3.2.9). ∎

Remarks

(a) Statement (d) of Proposition 3.2.2 also follows from (1.4.19) since $b_n > n/a_n \geq Cn^\varepsilon$. Note that estimates $n/a_n < b_n \leq (1+C_1)n/a_n$ are proved in [72, p. 78] by a different method. In particular, the integral representation

$$b_n = \int_0^n \frac{dz}{a_z}$$

is used (see Proposition 1.4.7(g)).

(b) For $W = W_\alpha$, $\alpha > 1$, we have $b_n a_n = n\alpha/(\alpha-1)$ (see (10.1.4)). This shows that the quantities $\liminf_{n\to\infty} b_n a_n/n$ and $\limsup_{n\to\infty} b_n a_n/n$ can take any values between 1 and ∞ (for Freud weights).

(c) Mhaskar [91, p. 245] used the constant b_n for a description of the strip of convergence for Freud expansions on $\mathbb{R}$.

PROPOSITION 3.2.3. *Let $W \in \mathcal{F}$. Then there exist constants $C_0 = C_0(W) \in (0,1)$ and $C = C(W)$ such that for all $u \in (0, C_0)$,*

$$\text{(3.2.13)} \qquad 1 - \frac{h(1-u)}{b_n} \leq Cu^{\frac{2(\Lambda-1)}{\Lambda+1}},$$

where Λ is defined in (1.4.4).

PROOF. We first establish an upper bound for $h(1) - h(1-u)$. Using (3.2.2) for $r \in [1/2, 1]$, we have

$$h(1) - h(r) = \frac{2}{\pi} \int_0^\pi Q'(a_n \sin t) \sin t \frac{(1-r)^2}{2(1-\cos t)(1+r^2-2r\cos t)} dt$$

$$= \frac{1}{\pi} \int_0^\pi Q'(a_n \sin t) \cot(t/2) \frac{(1-r)^2}{(1-r)^2 + 4r\sin^2(t/2)} dt$$

(3.2.14)
$$\leq C \int_0^\pi Q'(a_n \sin t) \cot(t/2) \frac{(1-r)^2}{(1-r)^2 + t^2} dt.$$

Since $\frac{\cot(t/2)}{(1-r)^2 + t^2}$ is decreasing in $t \in (0, \pi)$, we get

$$\int_0^\pi Q'(a_n \sin t) \cot(t/2) \frac{(1-r)^2}{(1-r)^2 + t^2} dt$$

$$= (1-r)^2 \int_0^{\pi/2} Q'(a_n \cos t) \left[\frac{\cot(\pi/4 - t/2)}{(1-r)^2 + (\pi/2 - t)^2} \right.$$

$$\left. + \frac{\cot(\pi/4 + t/2)}{(1-r)^2 + (\pi/2 + t)^2} \right] dt$$

$$\leq 2 \int_0^{\pi/2} Q'(a_n \sin t) \cot(t/2) \frac{(1-r)^2}{(1-r)^2 + t^2} dt$$

(3.2.15)
$$\leq 4 \int_0^{\pi/2} \frac{Q'(a_n \sin t)}{\sin t} \frac{(1-r)^2}{(1-r)^2 + t^2} dt.$$

Next, (3.2.14) and (3.2.15) imply the following estimate for $0 \leq u \leq 1/2$:

(3.2.16)
$$h(1) - h(1-u) \leq C \int_0^{\pi/2} \frac{Q'(a_n \sin t)}{\sin t} \frac{u^2}{u^2 + t^2} dt.$$

Further using (1.4.13), we obtain for $y \in (0, 1/2)$,

(3.2.17)
$$\int_0^y \frac{Q'(a_n \sin t)}{\sin t} \frac{u^2}{u^2 + t^2} dt \leq (\pi/2) \int_0^y \frac{Q'(a_n x)}{x} dx \leq CQ'(a_n y).$$

Furthermore for $y \in (0, 1/2)$,

(3.2.18)
$$\int_y^{\pi/2} \frac{Q'(a_n \sin t)}{\sin t} \frac{u^2}{u^2 + t^2} dt \leq \frac{u^2}{y^2} \int_0^{\pi/2} \frac{Q'(a_n \sin t)}{\sin t} dt.$$

Since by (3.2.4),

$$b_n = h(1) = \frac{2}{\pi} \int_0^{\pi/2} \frac{Q'(a_n \sin t)}{\sin t} dt \geq \frac{2}{\pi} \int_{\pi/6}^{\pi/2} \frac{Q'(a_n \sin t)}{\sin t} dt$$

(3.2.19)
$$\geq CQ'(a_n/2),$$

we obtain from (3.2.16)–(3.2.19)

$$1 - \frac{h(1-u)}{b_n} \leq C \left(\frac{Q'(a_n y)}{Q'(a_n/2)} + \frac{u^2}{y^2} \right),$$

where $u \in [0, 1/2]$, $y \in [0, 1/2)$ and C is independent of n, u, and y. Hence by (1.4.14) and by property (a) of Definition 1.4.3, we have

(3.2.20)
$$1 - \frac{h(1-u)}{b_n} \leq C \left(\frac{T(a_n y)}{T(a_n/2)} y^{\Lambda-1} + \frac{u^2}{y^2} \right) \leq C \left(y^{\Lambda-1} + \frac{u^2}{y^2} \right)$$

for all $u \in [0, 1/2]$, $y \in [0, 1/2]$. Choosing $y = u^{\frac{2}{1+\Lambda}}$ in (3.2.20), we have that for any $u \in \left(0, 2^{-\frac{1+\Lambda}{2}}\right)$, inequality (3.2.13) holds. So Proposition 3.2.3 is established with $C_0 = 2^{-\frac{1+\Lambda}{2}}$. ∎

PROPOSITION 3.2.4. *Let $W \in \mathcal{F}$. Then the following statements hold:*

(a) *For $r \in [0, 1]$,*

$$\frac{h(r)}{b_n} \geq \frac{2Cr}{(1+r)^2}. \tag{3.2.21}$$

where $C \in (0, 1)$.

(b) *If $\lim_{x \to \infty} T(x) = \infty$, then for $r \in [0, 1]$,*

$$\frac{h(r)}{b_n} \geq \frac{2r}{(1+r)^2}(1 + o(1)), \qquad n \to \infty, \tag{3.2.22}$$

where $o(1)$ is independent of r.

PROOF. Using (1.4.12), (3.2.2), and (3.2.5), we have

$$\frac{h(r)}{b_n} = \frac{2r}{\pi b_n} \int_0^\pi \frac{Q'(a_n \sin t) \sin t}{1 + r^2 + 2r \cos t} \, dt$$

$$\geq \frac{2r}{\pi(1+r)^2 b_n} \int_{-\pi/2}^{\pi/2} Q'(a_n \cos t) \cos t \, dt = \frac{2rn}{(1+r)^2 b_n a_n}$$

$$\geq \frac{2r}{(1+r)^2}\left(1 + \frac{C}{\sqrt{T(a_n)}}\right)^{-1}.$$

This yields (3.2.21) and (3.2.22). ∎

PROPOSITION 3.2.5. *If $W \in \mathcal{F}$ and $0 < C_0 < C_1 < 1$, then for all $r_1, r_2 \in [C_0, C_1]$ with $r_2 > r_1$,*

$$C_3(r_2 - r_1) \leq 1 - \frac{h(r_1)(1 + r_1^2)(1 - r_2^2)}{h(r_2)(1 + r_2^2)(1 - r_1^2)} \leq C_2(r_2 - r_1), \tag{3.2.23}$$

where C_2 and C_3 depend only on C_0, C_1, and W.

PROOF. Let us set

$$\varphi(r, t) := \frac{r(1 + r^2)}{(1 - r^2)(1 + r^2 + 2r \cos t)}, \qquad r \in (0, 1), \ t \in [0, \pi].$$

It is easy to verify the relations

$$\frac{1}{8} \leq \frac{1 + r^2}{(1+r)^4} \leq \frac{\partial \varphi(r, t)}{\partial r}$$

$$= \frac{4r^2}{(1-r^2)^2(1+r^2+2r\cos t)} + \frac{1+r^2}{(1+r^2+2r\cos t)^2}$$

$$\leq \frac{4r^2 + (1+r^2)(1+r)^2}{(1-r)^4(1+r)^2} \leq \frac{3}{(1-r)^4}, \qquad r \in (0, 1), \ t \in [0, \pi].$$

Hence

$$\frac{1}{8} \leq \min_{r\in[0,C_1]} \min_{t\in[0,\pi]} \frac{\partial \varphi(r,t)}{\partial r}$$

(3.2.24)
$$\leq \max_{r\in[0,C_1]} \max_{t\in[0,\pi]} \frac{\partial \varphi(r,t)}{\partial r} \leq \frac{3}{(1-C_1)^4}.$$

Next using (3.2.2), (3.2.24), and (1.4.12), we get

$$\frac{n(r_2-r_1)}{4a_n} \leq \frac{h(r_2)(1+r_2^2)}{1-r_1^2} - \frac{h(r_1)(1+r_1^2)}{1-r_1^2}$$

$$= \frac{2}{\pi}\int_0^\pi Q'(a_n \sin t)\sin t\big(\varphi(r_2,t)-\varphi(r_1,t)\big)\,dt$$

(3.2.25)
$$\leq \frac{6n(r_2-r_1)}{a_n(1-C_1)^4}.$$

Further making use of statements (c), (d) of Proposition 3.2.1 and taking account of (1.4.12), (3.2.6), we have

$$\frac{2C(W)n}{(1-C_1^2)a_n} \geq \frac{b_n(1+r_2^2)}{1-r_2^2} \geq \frac{h(r_2)(1+r_2^2)}{1-r_2^2}$$

(3.2.26)
$$\geq \frac{C_0}{2\pi}\int_0^\pi Q'(a_n\sin t)\sin t\,dt = \frac{C_0 n}{2a_n}.$$

Thus (3.2.23) follows from (3.2.25) and (3.2.26). ∎

PROPOSITION 3.2.6. *If $W \in \mathcal{F}_C$, then for all $\varphi \in [0,\pi/2]$,*

(3.2.27)
$$Q(a_n \cos\varphi) \leq (\pi/4)a_n b_n \cos\varphi.$$

PROOF. By (3.2.4), we have

(3.2.28)
$$(\pi/4)a_n b_n = a_n\int_0^{\pi/2}\frac{Q'(a_n\sin t)\cos t}{\sin 2t}\,dt \geq Q(a_n).$$

Since $Q(a_n y)$ is convex in $[0,1]$ and $Q(0)=0$, we obtain from (3.2.28)

$$Q(a_n\cos\varphi) \leq Q(a_n)\cos\varphi \leq (\pi/4)a_n b_n\cos\varphi.$$

Hence (3.2.27) holds. ∎

3.3. The Functions $\phi(r)$ and $\phi_1(r)$

We first study some properties of the function

(3.3.1)
$$\phi(r) = \phi(Q,n,r) := \frac{a_n h(r)}{2}\left(\frac{1}{r}-r\right) - H(ir) + n\log r.$$

PROPOSITION 3.3.1. *If $W \in \mathcal{F}_M$, then ϕ is increasing in $(0,1)$.*

PROOF. Differentiating (3.3.1) and taking account of (3.2.1) and (3.2.3), we have

$$\phi'(r) = \frac{a_n}{2}h'(r)\left(\frac{1}{r}-r\right) > 0, \qquad 0 < r < 1.$$

This proves the proposition. ∎

Since by (3.1.4) and (3.2.4), $\phi(1)=0$ (at least for $W\in\mathcal{F}_C$), Proposition 3.3.1 implies that $\phi(r) \leq 0$ for $r\in(0,1]$. However, we shall need a more precise estimate.

PROPOSITION 3.3.2. *If $W \in \mathcal{F}_M$, then for $0 \leq r < 1$,*

(3.3.2) $$\phi(r) \leq -(n/36)(1-r^2)^3 \leq -(n/36)(1-r)^3.$$

The proof is based on three lemmas in which we assume that $W \in \mathcal{F}_M$. We first establish an integral representation for ϕ.

LEMMA 3.3.3. *For $0 \leq r < 1$,*

(3.3.3) $$\phi(r) = \frac{2a_n}{\pi} \int_0^{\pi/2} Q'(a_n \cos t) \phi(r,t)\, dt,$$

where

(3.3.4) $$\phi(r,t) := -\sin t \arctan\left(\frac{1-r^2}{1+r^2}\tan t\right) + \cos t \log r + \frac{(1-r^4)\cos t}{1+r^4+2r^2\cos 2t}.$$

PROOF. Note first that by (3.1.17), (3.1.18), and (3.2.2),

$$h(r) = \frac{r}{\pi}\int_{-\pi}^{\pi}\frac{Q'(a_n\sin t)\sin t}{1+r^2+2r\cos t}\,dt = \frac{r}{\pi}\int_{-\pi}^{\pi}\frac{Q'(a_n\cos t)\cos t}{1+r^2+2r\sin t}\,dt$$

(3.3.5) $$= \frac{r}{\pi}\int_0^{\pi/2} Q'(a_n\cos t) A\left(\frac{\cos t}{1+r^2+2r\sin t}, t\right)\,dt,$$

where

$$A\left(\frac{\cos t}{1+r^2+2r\sin t}, t\right) = 2\cos t\left[\frac{1}{1+r^2+2r\sin t}+\frac{1}{1+r^2-2r\sin t}\right]$$

(3.3.6) $$= \frac{4\cos t(1+r^2)}{1+r^4+2r^2\cos 2t}.$$

Then (3.3.5) and (3.3.6) imply

(3.3.7) $$\frac{a_n h(r)}{2}\left(\frac{1}{r}-r\right) = \frac{2a_n}{\pi}\int_0^{\pi/2} Q'(a_n\cos t)\frac{(1-r^4)\cos t}{1+r^4+2r^2\cos 2t}\,dt.$$

Similarly by (1.4.12),

$$n\log r = \frac{a_n\log r}{2\pi}\int_{-\pi}^{\pi} Q'(a_n\cos t)\cos t\, dt$$

$$= \frac{a_n\log r}{2\pi}\int_0^{\pi/2} Q'(a_n\cos t) A(\cos t, t)\, dt$$

(3.3.8) $$= \frac{2a_n}{\pi}\int_0^{\pi/2} Q'(a_n\cos t)\log r\cos t\, dt.$$

Thus (3.3.3) follows from (3.1.22), (3.3.7), and (3.3.8). ∎

Some properties of the kernel $\phi(r,t)$ are given below.

LEMMA 3.3.4. (a) *For $0 < r < 1$ and $0 \leq t < \pi/2$,*

(3.3.9) $$\frac{\partial \phi(r,t)}{\partial r} = \frac{\cos t\,(1-r^2)^2(1+4r^2+r^4-2r^2\cos 2t)}{r(1+r^4+2r^2\cos 2t)} > 0.$$

(b) $\phi(1,t) = 0$, $t \in [0,\pi/2)$.
(c) *For $0 \leq r < 1$, $0 \leq t < \pi/2$,*

(3.3.10) $$\phi(r,t) < 0.$$

3.3. THE FUNCTIONS $\phi(r)$ AND $\phi_1(r)$

PROOF. By straightforward calculation,

$$\begin{aligned}\frac{\partial \phi(r,t)}{\partial r} &= \cos t \left[\frac{4\sin^2 t\, r}{1+r^4+2r^2\cos 2t} + \frac{\cos t}{r} - \frac{4r^3}{1+r^4+2r^2\cos 2t} \right. \\ &\quad \left. - \frac{(1-r^4)(4r^3+4r\cos 2t)}{(1+r^4+2r^2\cos 2t)^2} \right] \\ &= \frac{\cos t\,(1-r^2)[(1+r^4+2r^2\cos 2t)(1+3r^2)-(1+r^2)(4r^4+4r^2\cos 2t)]}{r(1+r^4+2r^2\cos 2t)^2} \\ &= \frac{\cos t\,(1-r^2)^2(1+4r^2+r^4-2r^2\cos 2t)}{r(1+r^4+2r^2\cos 2t)^2} > 0,\end{aligned}$$

if $r \in (0,1)$, $t \in [0, \pi/2)$. Thus (3.3.9) holds.

Since statement (b) immediately follows from (3.3.4), we obtain (3.3.10) from statements (a) and (b) in the case $0 < r < 1$ and $0 \leq t < \pi/2$. To prove statement (c), it remains to note that

$$\lim_{r \to 0+} \phi(r,t) = -\infty, \qquad t \in [0, \pi/2). \qquad \blacksquare$$

LEMMA 3.3.5. *For $0 \leq r \leq 1$ and $0 \leq t \leq \pi/3$,*

$$|\phi(r,t)| \geq \frac{\cos t\,(1-r^2)^3}{24}. \tag{3.3.11}$$

PROOF. We first prove the estimate

$$-\cos t \left(\log r + \frac{1-r^4}{(1+r^2)^2} \right) \geq \frac{\cos t\,(1-r^2)^3}{24}. \tag{3.3.12}$$

We have for $0 < r \leq 1$,

$$\begin{aligned}-\log r - \frac{1-r^4}{(1+r^2)^2} &= -\frac{1}{2}\log(1-(1-r^2)) - \frac{1-r^2}{2(1-(1-r^2)/2)} \\ &= \frac{1}{2}\sum_{k=1}^{\infty} \frac{(1-r^2)^k}{k} - \sum_{k=1}^{\infty} \left(\frac{1-r^2}{2}\right)^k \\ &= \frac{(1-r^2)^3}{2}\left(\frac{1}{12} + \sum_{k=1}^{\infty} \left(\frac{1}{k+3} - \frac{1}{2^{k+2}}\right)(1-r^2)^k\right) \\ &\geq \frac{(1-r^2)^3}{24}.\end{aligned}$$

This yields (3.3.12). Next using the elementary inequality

$$\arctan y \geq y - y^3/3, \quad y > 0,$$

we obtain for $0 < r < 1$ and $0 \le t < \pi/2$,

$$\begin{aligned}
\Delta(r,t) &:= \sin t \arctan\left(\frac{1-r^2}{1+r^2}\tan t\right) - \frac{(1-r^4)\cos t}{1+r^4+2r^2\cos 2t} \\
&\quad + \frac{(1-r^4)\cos t}{(1+r^2)^2} \\
&= \sin t \arctan\left(\frac{1-r^2}{1+r^2}\tan t\right) - \frac{4r^2\sin^2 t\cos t}{1+r^4+2r^2\cos 2t}\left(\frac{1-r^2}{1+r^2}\right) \\
&\ge \sin t\tan t\left(\frac{1-r^2}{1+r^2}\right) - \frac{1}{3}\sin t\tan^3 t\left(\frac{1-r^2}{1+r^2}\right)^3 \\
&\quad - \frac{4r^2\sin^2 t\cos t}{1+r^4+2r^2\cos 2t}\left(\frac{1-r^2}{1+r^2}\right) \\
&= \sin t\tan t\frac{(1-r^2)^2}{1+r^4+2r^2\cos 2t}\left(\frac{1-r^2}{1+r^2}\right) - \frac{1}{3}\sin t\tan^3 t\left(\frac{1-r^2}{1+r^2}\right)^3 \\
&\ge \sin t\tan t\left(1 - \frac{1}{3}\tan^2 t\right)\left(\frac{1-r^2}{1+r^2}\right)^3.
\end{aligned}$$

Hence

(3.3.13) $$\Delta(r,t) \ge 0, \quad 0 < r \le 1, \ 0 \le t \le \pi/3.$$

Finally, we obtain from (3.3.4), (3.3.10), (3.3.12), and (3.3.13)

$$|\phi(r,t)| = \Delta(r,t) - \cos t\left(\log r + \frac{1-r^4}{(1+r^2)^2}\right) \ge \frac{\cos t\,(1-r^2)^3}{24}.$$

This proves the lemma. ∎

Proof of Proposition 3.3.2. Using (3.3.3), (3.3.4), (3.3.10), and (3.3.11), we have for $0 \le r < 1$

$$\begin{aligned}
|\phi(r)| &= \frac{2a_n}{\pi}\int_0^{\pi/2} Q'(a_n\cos t)|\phi(r,t)|\,dt \\
&\ge \frac{2a_n}{\pi}\int_0^{\pi/3} Q'(a_n\cos t)|\phi(r,t)|\,dt \\
(3.3.14)\quad &\ge \frac{a_n(1-r^2)^3}{12\pi}\int_0^{\pi/3} Q'(a_n\cos t)\cos t\,dt.
\end{aligned}$$

Further by property (d) of Definition 1.4.1, the function $Q'(a_n\cos t)\cos t$ is decreasing on $[0,\pi/2]$, and this yields the inequality

(3.3.15) $$\int_0^{\pi/3} Q'(a_n\cos t)\cos t\,dt \ge \frac{2}{3}\int_0^{\pi/2} Q'(a_n\cos t)\cos t\,dt = \frac{\pi n}{3a_n}.$$

Thus (3.3.14) and (3.3.15) imply (3.3.2). ∎

The behavior of the function

(3.3.16) $$\phi_1(r) = \phi_1(Q,n,r) := \frac{a_n h(r)(1+r^2)^2}{2r(1-r^2)} - H(ir) + n\log r$$

is studied in the following two propositions:

3.3. THE FUNCTIONS $\phi(r)$ AND $\phi_1(r)$

PROPOSITION 3.3.6. *Let $W \in \mathcal{F}_M$. Then the following statements hold:*

(a) *For $0 < r < 1$,*
$$\phi_1'(r) = \frac{a_n}{2} h_1'(r) \left(\frac{1}{r} + r\right),$$
where
$$h_1(r) := h(r) \frac{1 + r^2}{1 - r^2}.$$

(b) *ϕ_1 is increasing in $(0, 1)$.*

(c)
$$\lim_{r \to 0+} \phi_1(r) = -\infty, \quad \lim_{r \to 1-} \phi_1(r) = \infty.$$

(d) *There exists the unique solution $r_0 = r_0(Q, n)$ to the equation*

(3.3.17)
$$\phi_1(r) = 0.$$

PROOF. Using (3.2.1), we have

$$\begin{aligned}
\phi_1'(r) &= \left(\frac{a_n}{2} h_1(r)\left(\frac{1}{r} + r\right) - H(ir) + n \log r\right)' \\
&= \frac{a_n}{2} h_1'(r)\left(\frac{1}{r} + r\right) - \frac{a_n h_1(r)(1 - r^2)}{2r^2} + \frac{a_n h(r)(1 + r^2)}{2r^2} \\
&= \frac{a_n}{2} h_1'(r)\left(\frac{1}{r} + r\right).
\end{aligned}$$

Hence statement (a) follows. Since h_1 is increasing in $(0, 1)$, statement (a) implies (b). Next by (3.1.3) and (3.2.2), the limit

$$\lim_{r \to 0+} \left(\frac{a_n h(r)(1 + r^2)^2}{2r(1 - r^2)} - H(ir)\right)$$

exists. Therefore, $\lim_{r \to 0+} \phi_1(r) = -\infty$. Further, it follows from (3.2.2) that

$$\inf_{r \in (0,1]} h(r)/r \geq \frac{1}{2\pi} \int_0^\pi Q'(a_n \sin t) \sin t \, dt > 0.$$

This implies $\lim_{r \to 1-} \phi_1(r) = \infty$. Statement (d) is an immediate consequence of (b) and (c). ∎

It follows from Proposition 3.3.6 that $\phi_1(r) < 0$ for $0 < r < r_0$. A more precise estimate is given below.

PROPOSITION 3.3.7. *If $W \in \mathcal{F}_M$, then for $0 \leq r \leq r_0$,*

(3.3.18)
$$\phi_1(r) \leq -n(r_0 - r),$$

where r_0 is defined in Proposition 3.3.6(d).

PROOF. We first note that by (3.1.22), (3.3.7), and (3.3.8),

(3.3.19)
$$\phi_1(r) = \frac{2a_n}{\pi} \int_0^{\pi/2} Q'(a_n \cos t) \phi_1(r, t) \, dt, \quad 0 < r < 1,$$

where

$$\phi_1(r,t) := -\sin t \arctan\left(\frac{1-r^2}{1+r^2}\tan t\right) + \cos t \log r$$

(3.3.20)
$$+ \frac{(1+r^2)^3 \cos t}{(1-r^2)(1+r^4+2r^2\cos 2t)}.$$

Next by a straightforward calculation,

(3.3.21) $\quad \dfrac{\partial}{\partial r}\left(\dfrac{(1+r^2)^3}{(1-r^2)(1+r^4+2r^2\cos 2t)}\right)$

$$= \frac{4r(1+r^2)^2[(2-\cos 2t)r^4 - 2(1-2\cos 2t)r^2 + 2 - \cos 2t]}{(1-r^2)^2(1+r^4+2r^2\cos 2t)^2} > 0,$$

if $0 < r < 1$ and $0 \leq t \leq \pi/2$. Then (3.3.21) shows that all three summands in the right-hand side of (3.3.20) are increasing in $r \in (0,1)$. Hence the following inequality holds for $0 < r < r_0$

(3.3.22) $\quad \phi_1(r_0,t) - \phi_1(r,t) > \cos t(\log r_0 - \log r) > \cos t(r_0 - r).$

Further using (1.4.12), (3.3.19), (3.3.22) and taking account of the equality $\phi_1(r_0) = 0$, we have

$$\phi_1(r) = \frac{2a_n}{\pi}\int_0^{\pi/2} Q'(a_n \cos t)\bigl(\phi_1(r,t) - \phi_1(r_0,t)\bigr)\,dt$$

$$\leq -\frac{2a_n(r_0-r)}{\pi}\int_0^{\pi/2} Q'(a_n \cos t)\cos t\,dt = -n(r_0 - r).$$

Hence (3.3.18) follows. ∎

Finally, we establish numerical estimates for r_0 that are independent of Q and n.

PROPOSITION 3.3.8. *If $W \in \mathcal{F}_M$, then the following inequalities hold:*

(3.3.23) $\quad\quad\quad\quad\quad\quad\quad r_1 < r_0 < r_2,$

where

(3.3.24) $\quad\quad\quad\quad r_1 = 0.24166, \quad r_2 = 0.59418.$

PROOF. Note first that the equation

$$\ell(r) := A + B\log r + \frac{C(1+r^2)^k}{(1-r^2)^s} = 0$$

has the only solution in $(0,1)$ for $A \in \mathbb{R}$, $B > 0$, $C > 0$, $k > 0$, and $s > 0$, since $\ell(r)$ is increasing in $r \in (0,1)$ and $\ell(0+) = -\infty$, $\ell(1-) = \infty$.

Next by (3.3.19) and (3.3.20),

(3.3.25) $\quad 0 = \dfrac{2a_n}{\pi}\displaystyle\int_0^{\pi/2} Q'(a_n\cos t)\phi_1(r_0,t)\,dt$

$$\geq \frac{2a_n}{\pi}\int_0^{\pi/2} Q'(a_n\cos t)\left(-\frac{\pi}{2}\sin t + \cos t\log r_0 + \frac{1+r_0^2}{1-r_0^2}\cos t\right)dt.$$

Further, we need the inequality

$$\int_0^{\pi/2} Q'(a_n \cos t) \sin t \, dt \leq \int_0^{\pi/2} Q'(a_n \cos t) \cos t \, dt \qquad (3.3.26)$$

that can be deduced from the following inequality for continuous functions f and g on $[0, \pi/2]$ (see [25, p. 22]):

$$\int_0^{\pi/2} fg \, dt \leq \int_0^{\pi/2} f^* g^* \, dt, \qquad (3.3.27)$$

where f^* and g^* are the decreasing rearrangements of f and g on $[0, \pi/2]$. Since

$$[Q'(a_n \cos \cdot)]^*(t) = Q'(a_n \cos t), \quad [\sin \cdot]^*(t) = \cos t, \quad t \in [0, \pi/2], \qquad (3.3.28)$$

inequality (3.3.26) immediately follows from (3.3.27) and (3.3.28).

Then by (3.3.25) and (3.3.26),

$$-\frac{\pi}{2} + \log r_0 + \frac{1 + r_0^2}{1 - r_0^2} \leq 0.$$

Thus $r_0 \leq r_2^*$, where $r_2^* \in (0, 1)$ is the solution to the equation

$$-\frac{\pi}{2} + \log r + \frac{1 + r^2}{1 - r^2} = 0. \qquad (3.3.29)$$

Similarly we obtain the lower estimate for r_0,

$$0 \leq \frac{2 a_n}{\pi} \int_0^{\pi/2} Q'(a_n \cos t) \left(\cos t \log r_0 + \left(\frac{1 + r_0^2}{1 - r_0^2} \right)^3 \cos t \right) dt.$$

This implies $r_1^* \leq r_0$, where $r_1^* \in (0, 1)$ is the solution to the equation

$$\log r + \left(\frac{1 + r^2}{1 - r^2} \right)^3 = 0. \qquad (3.3.30)$$

Finally finding the numerical solutions to equations (3.3.29) and (3.3.30), we arrive at the estimates $r_1 < r_1^* \leq r_0 \leq r_2^* < r_2$. ∎

3.4. The Main Estimate for Re $H(w)$

The following estimate is the chief ingredient in the proof of Theorem 2.3.1:

THEOREM 3.4.1. *If $W \in \mathcal{F}_C$, then for every $w \in \mathbb{C}$, satisfying $0 < |w| \leq 1$, we have*

$$\operatorname{Re} H(w) \leq \frac{a_n b_n}{2} \left| w + \frac{1}{w} \right| + n \log |w|. \qquad (3.4.1)$$

PROOF. Since the function $-\frac{a_n b_n}{2} \left| w + \frac{1}{w} \right|$ is not subharmonic for $0 < |w| < 1$, we shall replace it with a piecewise subharmonic function $-m(w) \geq -\frac{a_n b_n}{2} \left| w + \frac{1}{w} \right|$.

Let us set for $w = re^{i\varphi}$,

$$m(w) = M(r, \varphi)$$

(3.4.2)
$$:= \frac{a_n b_n}{2} \max\left(\left|\operatorname{Re}\left(w + \frac{1}{w}\right)\right|, \left|\operatorname{Im}\left(w + \frac{1}{w}\right)\right|, \right.$$
$$\left. \sqrt{2\left|\operatorname{Im}\left(w + \frac{1}{w}\right)\operatorname{Re}\left(w + \frac{1}{w}\right)\right|}\right)$$
$$= \frac{a_n b_n}{2}\left(\left(\frac{1}{r} + r\right)|\cos\varphi|, \left(\frac{1}{r} - r\right)|\sin\varphi|, \sqrt{\left(\frac{1}{r^2} - r^2\right)|\sin 2\varphi|}\right).$$

It is easy to verify the inequality

(3.4.3) $$m(w) \le \frac{a_n b_n}{2}\left|w + \frac{1}{w}\right|, \quad 0 < |w| \le 1.$$

Thus (3.4.1) follows from (3.4.2), (3.4.3) and the estimate

(3.4.4) $$G(r, \varphi) := \operatorname{Re} H(re^{i\varphi}) - n\log r - M(r, \varphi) \le 0, \quad 0 < r \le 1, \; |\varphi| \le \pi.$$

To prove (3.4.4), we first note that $M(r, \varphi)$ and $\operatorname{Re} H(re^{i\varphi})$ are even and π-periodic functions in $\varphi \in [-\pi, \pi]$ for each $r \in (0, 1]$. Indeed, this statement for $M(r, \varphi)$ follows from (3.4.2), while for $\operatorname{Re} H(re^{i\varphi})$, it is proved in Proposition 3.1.1(d).

Thus $G(r, \varphi)$ is even and periodic in φ for each $r \in (0, 1]$. Therefore it suffices to prove the inequality

(3.4.5) $$G(r, \varphi) \le 0, \quad 0 < r \le 1, \; 0 \le \varphi \le \pi/2,$$

that implies (3.4.4).

To prove (3.4.5), we consider two curves K_1, K_2 and three domains D_1, D_2, D_3 located in the first quadrant

$$B_+ := \{(r, \varphi) : 0 \le r \le 1, \; 0 \le \varphi \le \pi/2\}$$

of the unit disk. They are defined below in terms of polar coordinates (r, φ).

$$K_1 := \left\{(r, \varphi) : r^2 = \frac{\tan\varphi - 2}{\tan\varphi + 2}, \; \arctan 2 \le \varphi \le \pi/2\right\},$$

$$K_2 := \left\{(r, \varphi) : r^2 = \frac{\tan\varphi - 1/2}{\tan\varphi + 1/2}, \; \arctan 1/2 \le \varphi \le \pi/2\right\},$$

$$D_1 := \left\{(r, \varphi) : \frac{\tan\varphi - 1/2}{\tan\varphi + 1/2} < r^2 < 1, \; 0 < \varphi < \pi/2\right\},$$

$$D_2 := \left\{(r, \varphi) : \frac{\tan\varphi - 2}{\tan\varphi + 2} < r^2 < \frac{\tan\varphi - 1/2}{\tan\varphi + 1/2}, \; 0 < \varphi < \pi/2\right\},$$

$$D_3 := \left\{(r, \varphi) : 0 < r^2 < \frac{\tan\varphi - 2}{\tan\varphi + 2}, \; 0 < \varphi < \pi/2\right\}.$$

It is easy to see that K_1 and K_2 are continuous curves with starting points at the origin and ending points at $(1, \pi/2)$. Also, D_1, D_2, and D_3 are mutually disjoint domains with the boundaries

(3.4.6) $$\partial D_1 = K_2 \cup \Gamma_1 \cup \Gamma_2, \quad \partial D_2 = K_1 \cup K_2, \quad \partial D_3 = K_1 \cup \Gamma_3,$$

where
$$\Gamma_1 := \{(1,\varphi): 0 \le \varphi \le \pi/2\}, \quad \Gamma_2 := \{(r,0): 0 \le r \le 1\},$$
$$\Gamma_3 := \{(r,\pi/2): 0 \le r \le 1\}.$$

Also we have

(3.4.7) $$B_+ = \overline{D}_1 \cup \overline{D}_2 \cup \overline{D}_3.$$

∎

Next we establish some properties of $M(r,\varphi)$ and $G(r,\varphi)$.

LEMMA 3.4.2. (a) *The following representation holds:*

(3.4.8) $$\frac{2M(r,\varphi)}{a_n b_n} = \left(\frac{1}{r}+r\right)\cos\varphi, \quad (r,\varphi) \in D_1 \cup \Gamma_1 \cup \Gamma_2,$$

$$\frac{2M(r,\varphi)}{a_n b_n} = \left(\frac{1}{r}+r\right)\cos\varphi = \sqrt{\left(\frac{1}{r^2}-r^2\right)\sin 2\varphi},$$
(3.4.9) $$(r,\varphi) \in K_2,$$

(3.4.10) $$\frac{2M(r,\varphi)}{a_n b_n} = \sqrt{\left(\frac{1}{r^2}-r^2\right)\sin 2\varphi}, \quad (r,\varphi) \in D_2,$$

$$\frac{2M(r,\varphi)}{a_n b_n} = \left(\frac{1}{r}-r\right)\sin\varphi = \sqrt{\left(\frac{1}{r^2}-r^2\right)\sin 2\varphi},$$
(3.4.11) $$(r,\varphi) \in K_1,$$

(3.4.12) $$\frac{2M(r,\varphi)}{a_n b_n} = \left(\frac{1}{r}-r\right)\sin\varphi, \quad (r,\varphi) \in D_3 \cup \Gamma_3.$$

(b) *The function* $-M(r,\varphi)$ *is continuous everywhere in* B_+ *but the origin* O, *harmonic in* D_1, D_3, *and subharmonic in* D_2.

(c)

(3.4.13) $$\sup_{(r,\varphi)\in B_+} G(r,\varphi) = \sup_{\Gamma} G(r,\varphi),$$

where
$$\Gamma = K_1 \cup K_2 \cup \Gamma_1 \cup \Gamma_3 \cup \{O\}$$
and G is defined at the origin as $\limsup_{B_+ \ni (r,\varphi) \to O} G(r,\varphi).$

PROOF. We first note that $M(r,\varphi)$ is continuous in $B_+ \setminus \{O\}$. This immediately follows from the definition of M given in (3.4.2).

Next we show that the following statements hold:

(i) For $(r,\varphi) \in D_1$,

(3.4.14) $$\left(\frac{1}{r}+r\right)\cos\varphi > \sqrt{\left(\frac{1}{r^2}-r^2\right)\sin 2\varphi} > \left(\frac{1}{r}-r\right)\sin\varphi.$$

(ii) For $(r,\varphi) \in D_2$,

(3.4.15) $$\sqrt{\left(\frac{1}{r^2}-r^2\right)\sin 2\varphi} > \max\left(\left(\frac{1}{r}+r\right)\cos\varphi, \left(\frac{1}{r}-r\right)\sin\varphi\right).$$

(iii) For $(r, \varphi) \in D_3$,

(3.4.16) $\qquad \left(\dfrac{1}{r} - r\right) \sin \varphi > \sqrt{\left(\dfrac{1}{r^2} - r^2\right) \sin 2\varphi} > \left(\dfrac{1}{r} + r\right) \cos \varphi.$

It is easy to verify that the inequalities

$$\sqrt{\left(\dfrac{1}{r^2} - r^2\right) \sin 2\varphi} > \left(\dfrac{1}{r} - r\right) \sin \varphi, \qquad 0 < r < 1, \quad 0 < \varphi < \dfrac{\pi}{2}$$

are equivalent to

(3.4.17) $\qquad \dfrac{\tan \varphi - 2}{\tan \varphi + 2} < r^2 < 1, \qquad 0 < \varphi < \pi/2,$

and the inequalities

$$\left(\dfrac{1}{r} + r\right) \cos \varphi > \sqrt{\left(\dfrac{1}{r^2} - r^2\right) \sin 2\varphi}, \qquad 0 < r < 1, \quad 0 < \varphi < \pi/2,$$

are equivalent to

(3.4.18) $\qquad \dfrac{\tan \varphi - 1/2}{\tan \varphi + 1/2} < r^2 < 1, \qquad 0 < \varphi < \pi/2.$

Since the statements $(r, \varphi) \in D_1$ is equivalent to (3.4.18) and (3.4.18) implies (3.4.17), we obtain (3.4.14). Statements (ii) and (iii) can be proved similarly. Then (3.4.8), (3.4.10), and (3.4.12) immediately follow from (3.4.14), (3.4.15), and (3.4.16), respectively. Finally, (3.4.9) and (3.4.11) can be deduced from continuity of $M(r, \varphi)$ in $B_+ \setminus \{O\}$ and relations (3.4.6), (3.4.8), (3.4.10), and (3.4.12). Thus statement (a) of Lemma 3.4.2 follows.

Next note that the function

$$-\left(\dfrac{1}{r} - r\right) \sin \varphi = \operatorname{Im}\left(w + \dfrac{1}{w}\right), \qquad w = re^{i\varphi},$$

is harmonic in the set $\mathring{B}_+$ of all interior points of B_+, and the function

(3.4.19) $\qquad -\left(\dfrac{1}{r} + r\right) \cos \varphi = \operatorname{Re}\left(-w - \dfrac{1}{w}\right), \qquad w = re^{i\varphi},$

is harmonic in the semidisk $B_+^* := \{(r, \varphi) : 0 < r < 1, |\varphi| < \pi/2\}$. Also the function

$$\mu(w) := \left(\dfrac{1}{r^2} - r^2\right) \sin 2\varphi = \operatorname{Im}\left(-w^2 - \dfrac{1}{w^2}\right), \qquad w = re^{i\varphi},$$

is nonnegative and harmonic in $\mathring{B}_+$. Then for every $w = re^{i\varphi} \in \mathring{B}_+$ and any λ satisfying $0 < \lambda < \operatorname{dist}(w, \partial B_+)$, we have

$$\mu(w) = \dfrac{1}{2\pi} \int_0^{2\pi} \mu(w + \lambda e^{it})\, dt.$$

Applying the Schwarz inequality, we have

$$\sqrt{\mu(w)} \geq \dfrac{1}{2\pi} \int_0^{2\pi} \sqrt{\mu(w + \lambda e^{it})}\, dt.$$

This shows that the function $-\sqrt{\mu(w)} = -\sqrt{\left(\dfrac{1}{r^2} - r^2\right) \sin 2\varphi}$ is subharmonic in $\mathring{B}_+$. Thus taking into account (3.4.8), (3.4.10), and (3.4.12), we conclude that

$-M(r,\varphi)$ is harmonic in D_1, D_3 and subharmonic in D_2. Moreover, since $M(r,\varphi)$ is even in φ, we have from (3.4.8) and (3.4.19) that $-M(r,\varphi)$ is harmonic in

$$D_1^* = \left\{(r,\varphi) \colon \frac{|\tan\varphi| - 1/2}{|\tan\varphi| + 1/2} < r^2 < 1,\ |\varphi| < \pi/2\right\}.$$

Therefore the function $G(r,\varphi)$ is harmonic in D_1^*, D_3 and subharmonic in D_2. Applying now the maximum modulus principle for harmonic functions [86], we have

(3.4.20) $$\sup_{(r,\varphi)\in D_3} G(r,\varphi) = \sup_{(r,\varphi)\in K_1\cup\Gamma_3} G(r,\varphi),$$

and since $G(r,\varphi)$ is an even function in φ,

(3.4.21) $$\sup_{(r,\varphi)\in D_1} G(r,\varphi) = \sup_{(r,\varphi)\in D_1^*} G(r,\varphi) = \sup_{(r,\varphi)\in K_2\cup\Gamma_1} G(r,\varphi).$$

Next by the maximum modulus principle for subharmonic functions [86],

(3.4.22) $$\sup_{(r,\varphi)\in D_2} G(r,\varphi) = \sup_{(r,\varphi)\in K_1\cup K_2} G(r,\varphi).$$

Thus (3.4.13) follows from (3.4.20)–(3.4.22) and (3.4.7). ∎

Next, we need to estimate $G(r,\varphi)$ on Γ.

LEMMA 3.4.3. *If $W \in \mathcal{F}_c$, then*

(3.4.23) $$\sup_{(r,\varphi)\in\Gamma} G(r,\varphi) \leq 0.$$

PROOF. We study the behavior of $G(r,\varphi)$ on the following components of Γ:

(a) the origin O,
(b) the arc Γ_1,
(c) the interval Γ_3,
(d) the curves K_1 and K_2.

(a) *The origin.* We first establish lower estimates for $M(r,\varphi)$ in a neighborhood of the origin. If $(r,w) \in \overline{D}_1$ and $0 < r < 1/2$, then by (3.4.8) and (3.4.9),

(3.4.24) $$\frac{2M(r,\varphi)}{a_n b_n} \geq \frac{1}{r\sqrt{1+\tan^2\varphi}} \geq \frac{1}{r\sqrt{1+\frac{1}{4}\left(\frac{1+r^2}{1-r^2}\right)^2}} > \frac{0.7}{r}.$$

If $(r,w) \in \overline{D}_2$ and $0 < r < \frac{1}{2}$, then by (3.4.9), (3.4.10), and (3.4.11),

(3.4.25) $$\frac{2M(r,\varphi)}{a_n b_n} \geq \sqrt{2\left(\frac{1}{r^2}-r^2\right)\frac{\tan\varphi}{1+\tan^2\varphi}} \geq \frac{1+r^2}{r\sqrt{1+4\left(\frac{1+r^2}{1-r^2}\right)^2}} > \frac{0.2}{r}.$$

If $(r,w) \in \overline{D}_3$ and $0 < r < 1/2$, then by (3.4.11) and (3.4.12),

(3.4.26) $$\frac{2M(r,\varphi)}{a_n b_n} \geq \left(\frac{1}{r}-r\right)\sin\varphi \geq \left(\frac{1}{r}-r\right)\sin(\arctan 2) > \frac{0.6}{r}.$$

Then (3.4.24)–(3.4.26) imply

$$M(r,\varphi) \geq \frac{a_n b_n}{10r},\quad (r,\varphi) \in B_+,\ 0 < r < 1/2.$$

Hence
$$\limsup_{B_+ \ni (r,\varphi) \to O} G(r,\varphi) \leq \frac{1}{2\pi} \int_{-\pi}^{\pi} Q(a_n \cos t)\, dt + \limsup_{r \to 0} \left(n \log \frac{1}{r} - \frac{a_n b_n}{10r} \right)$$
(3.4.27)
$$= -\infty.$$

(b) *The arc* Γ_1. By Proposition 3.1.1(c) and relation (3.4.8), the restriction of $G(r,\varphi)$ to Γ_1 is

(3.4.28) $\qquad G(1,\varphi) = Q(a_n \cos \varphi) - a_n b_n \cos \varphi, \quad 0 \leq \varphi \leq \pi/2.$

Then the inequality
$$\sup_{(r,\varphi) \in \Gamma_1} G(r,\varphi) \leq 0$$
follows from Proposition 3.2.6.

(c) *The interval* Γ_3. By Proposition 3.1.2(a) and relation (3.4.12), we have
$$G(r,\pi/2) = H(ir) - n \log r - \frac{a_n b_n}{2}\left(\frac{1}{r} - r\right), \quad 0 < r \leq 1.$$

Hence
$$\frac{dG(r,\pi/2)}{dr} = \frac{a_n(1+r^2)}{2r^2}\left(\frac{2r(riH'(ir) - n)}{a_n(1+r^2)} + b_n\right)$$
(3.4.29)
$$= \frac{a_n(1+r^2)}{2r^2}(-h(r) + b_n),$$

where h is defined in (3.2.1). Statements (c) and (d) of Proposition 3.2.1 show that h is increasing in $[0,1)$ and $\lim_{r \to 1^-} h(r) = b_n$. So (3.4.29) implies that $G(r,\pi/2)$ is increasing in $(0,1)$. Since by (3.1.4), $H(i) = 0$, we have $G(1,\pi/2) = 0$. Therefore,

(3.4.30) $$\sup_{(r,\varphi) \in \Gamma_3} G(r,\varphi) \leq 0.$$

(d) *The curves* K_1 *and* K_2. We first note that identities (3.1.11), (3.2.4), and (3.3.8) imply the following integral representation for $G(r,\varphi)$:

(3.4.31) $$G(r,\varphi) = \frac{a_n}{\pi} \int_0^{\pi/2} Q'(a_n \cos t) F(r,t,\varphi)\, dt,$$

where
$$F(r,t,\varphi) := -\left(\sin t\, K(r,t,\varphi) + 2\cos t \log r + \frac{2M(r,\varphi)}{a_n b_n \cos t}\right).$$

Here, $K(r,t,\varphi)$ is the kernel defined in (3.1.12). In particular, by (3.4.9) and (3.4.11),
(3.4.32)
$$F(r,t,\varphi) = -\left(\sin t\, K(r,t,\varphi) + 2\cos t \log r + \begin{cases} \left(\frac{1}{r} - r\right)\frac{\sin \varphi}{\cos t}, & (r,\varphi) \in K_1 \\ \left(\frac{1}{r} + r\right)\frac{\cos \varphi}{\cos t}, & (r,\varphi) \in K_2 \end{cases}\right).$$

Then the estimate

(3.4.33) $$\sup_{(r,\varphi) \in K_1 \cup K_2} G(r,\varphi) \leq 0$$

follows from (3.4.31) and the inequality

(3.4.34) $\qquad F(r,t,\varphi) \leq 0, \quad 0 \leq t < \pi/2, \quad 0 \leq \varphi \leq \pi/2, \quad (r,\varphi) \in K_1 \cup K_2.$

The proof of this inequality is elementary though fairly long and technical. That is why the proof of (3.4.34) is given in Proposition A.1 of the Appendix.

Thus inequalities (3.4.27), (3.4.30), and (3.4.33) yield (3.4.23). ∎

We are now in position to complete the proof of Theorem 3.4.1 since (3.4.5) is an immediate consequence of estimates (3.4.13) and (3.4.23). ∎

CHAPTER 4

Polynomial Inequalities with Exponential Weights

In this chapter, we discuss Nikolskii-type inequalities in weighted metrics, prove new estimates for extremal polynomials, including the orthonormal polynomials associated with the weight W_1, and also estimate the growth of a polynomial in the complex plane. The proofs of Theorems 2.3.1 and 2.3.2 are given as well.

4.1. Nikolskii-type Inequalities

Various versions of the Nikolskii inequality for algebraic polynomials in weighted metrics were discussed in [72, 84, 91, 110], see these books for further references and detailed discussions.

In particular, Levin and Lubinsky [72, p. 295] proved that for $W \in \mathcal{F}(C^2)$ and $P \in \mathcal{P}_n$, the following inequality holds:

$$(4.1.1) \qquad \|P\|_{L_{q,W}(I)} \le CN(q,p,n)\|P\|_{L_{p,W}(I)},$$

where $0 < q, p \le \infty$ and

$$(4.1.2) \qquad N(q,p,n) := \begin{cases} a_n^{1/q-1/p}, & p \ge q \\ \left((n/a_n)\sqrt{T(a_n)}\right)^{1/p-1/q}, & p < q. \end{cases}$$

For a more general weight $W(x) = e^{-Q(x)}$, where Q is an even, continuous in $\mathbb{R}$ and increasing in $(0, \infty)$ function, Lubinsky and Saff [84, p. 53] established the following crude Nikolskii inequality:

$$(4.1.3) \qquad \|P\|_{L_{\infty,W}(\mathbb{R})} \le Cn^{2/\min(1,p)}\|P\|_{L_{p,W}(\mathbb{R})},$$

where $P \in \mathcal{P}_n$ and $0 < p < \infty$.

The following pointwise version of (4.1.1) was obtained in [72, p. 256]: for $W \in \mathcal{F}(C^2)$ there exist constants C and n_0 such that for any $P \in \mathcal{P}_n$, $n > n_0$ and for every $x \in I$,

$$(4.1.4) \qquad |P_n(x)|W(x) \le \left(C/\varphi_n(x)\right)^{1/p}\|P\|_{L_{p,W}(I)}, \qquad 0 < p < \infty,$$

where

$$(4.1.5) \qquad \varphi_n(x) := \begin{cases} \dfrac{|x^2 - a_{2n}^2|}{n\sqrt{(|x+a_n|+a_n\eta_n)(|x-a_n|+a_n\eta_n)}}, & |x| \le a_n, \\ \varphi_n(a_n), & |x| > a_n \end{cases}$$

and

$$(4.1.6) \qquad \eta_n := \left(nT(a_n)\right)^{-2/3}.$$

We shall need three corollaries of estimates (4.1.1), (4.1.3), and (4.1.4). We first establish two crude Nikolskii-type inequalities.

PROPOSITION 4.1.1. (a) If $W \in \mathcal{F}(C^2)$, then for any $P \in \mathcal{P}_n$, $n \geq 1$, and $0 < q, p \leq \infty$,

$$\|P\|_{L_{q,W}(I)} \leq C n^{2|1/p - 1/q|} \|P\|_{L_{p,W}(I)}. \tag{4.1.7}$$

(b) If $W \in \mathcal{F}(C^3)$, then for any $P \in \mathcal{P}_n$, $n \geq 1$, and $0 < q, p \leq \infty$,

$$\|P\|_{L_{q,W}(\mathbb{R})} \leq C n^{1/q + 2/\min(1,p)} \|P\|_{L_{p,W}(\mathbb{R})}. \tag{4.1.8}$$

PROOF. (a) Using (1.4.19) and (1.4.20), we have
$$C \leq a_n \leq C_1 n, \qquad T(a_n) \leq C n^2.$$
Hence the constant $N(p, q, n)$ defined by (4.1.2) satisfies the inequality

$$N(q, p, n) \leq C n^{2|1/p - 1/q|}. \tag{4.1.9}$$

Thus (4.1.1) and (4.1.9) yield (4.1.7).

(b) If $W \in \mathcal{F}(C^3)$, then (4.1.3) holds. Let us denote
$$\rho_n := 1 + C(\log n / n)^{2/3}.$$
Then using the L_q infinite-finite range inequality for $W \in \mathcal{F}(C^3)$ [84, Theorem 7.27] and taking account of the estimate $a_n \leq Cn$ that easily follows from (1.4.12), we have from (4.1.3)
$$\|P\|_{L_{q,W}(\mathbb{R})} \leq C\|P\|_{L_{q,W}[-\rho_n a_n, \rho_n a_n]}$$
$$\leq C\|P\|_{L_{\infty,W}(\mathbb{R})} (2\rho_n a_n)^{1/q} \leq C n^{1/q + 2/\min(1,p)} \|P\|_{L_{p,W}(\mathbb{R})}.$$
This proves the proposition. ∎

A special version of the Nikolskii inequality is presented below.

PROPOSITION 4.1.2. If $W \in \mathcal{F}(C^2)$, then for every $\delta \in (0,1)$ there exists a constant C such that for any $P \in \mathcal{P}_n$ and $0 < p \leq \infty$,

$$\sup_{x \in I} |P((1-\delta)x)| W(x) \leq C(n/a_n)^{1/p} \|P\|_{L_{p,W}(I)}. \tag{4.1.10}$$

PROOF. By (1.4.27), we have for any $P \in \mathcal{P}_n$ and every $\delta \in (0,1)$ that

$$\sup_{x \in I} |P((1-\delta)x)| W(x) = \max_{x \in [-a_n, a_n]} |P((1-\delta)x)| W(x). \tag{4.1.11}$$

Next using (4.1.4), we obtain for $|x| \leq a_n$
$$|P((1-\delta)x)| W(x) \leq |P((1-\delta)x)| W((1-\delta)x)$$
$$\leq (C/\varphi_n((1-\delta)x))^{1/p} \|P\|_{L_{p,W}(I)}. \tag{4.1.12}$$

Then (4.1.11) and (4.1.12) imply the estimate

$$\sup_{x \in I} |P((1-\delta)x)| W(x) \leq \left(C / \min_{|y| \leq (1-\delta) a_n} \varphi_n(y) \right)^{1/p} \|P\|_{L_{p,W}(I)}. \tag{4.1.13}$$

Since $a_{2n} > a_n$, we have from (4.1.5) and (4.1.6)

$$\left(\min_{|y| \leq (1-\delta) a_n} \varphi_n(y) \right)^{-1} \leq \frac{n a_n \sqrt{(2 - \delta + o(1))(1 + o(1))}}{a_{2n}^2 - a_n^2 (1-\delta)^2} \leq \frac{Cn}{a_n}, \tag{4.1.14}$$

as $n \to \infty$. Thus (4.1.13) and (4.1.14) yield (4.1.10). ∎

4.2. Extremal Polynomials

In this section we study properties of an nth L_q extremal polynomial
$$T_{n,q}(x) = T_{n,q}(W,x) := x^n + \cdots$$
of degree n, satisfying for $0 < q \leq \infty$
$$(4.2.1) \qquad E_{n,q}(W,I) := \inf_{P_{n-1} \in \mathcal{P}_{n-1}} \|x^n - P_{n-1}\|_{L_{q,W}(I)} = \|T_{n,q}\|_{L_{q,W}(I)}.$$

We first discuss some known properties of zeros of $T_{n,q}$.

PROPOSITION 4.2.1. *Let $W \in \mathcal{F}_M$ and let $1 < q < \infty$. Then the following statements hold:*

(a) *All zeros of $T_{n,q}$ are real, simple and belong to I, $n = 1, 2, \ldots$.*
(b) *If $W \in \mathcal{F}_C$, then the zeros $x_{k,n}^{(q)}$, $1 \leq k \leq n$, $n = 1, 2, \ldots$, of $T_{n,q}$ satisfy the inequalities*
$$(4.2.2) \qquad -a_{n+1/q} < x_{n,n}^{(q)} < \cdots < x_{1,n}^{(q)} < a_{n+1/q}.$$
(c) *If $W \in \mathcal{F}_C$, then all zeros of $T_{n,q}$ are located in the interval $\bigl[-a_n(1+C/(qn)), a_n(1+C/(qn))\bigr]$.*
(d) *If $W \in \mathcal{F}(C^3)$, then all zeros of $T_{n,q}$ are located in the interval $\bigl[-a_n(1+C(\log n/n)^{2/3}), a_n(1+C(\log n/n)^{2/3})\bigr]$.*

Statement (a) is a well-known fact in approximation theory (cf. [72, p. 314] and [84, p. 26]). Inequalities (4.2.2) are proved in [72, p. 314], and statement (c) follows from (4.2.2) and (1.4.23). The proof of statement (d) can be found in [84, Theorem 8.1].

Let us define the normalized extremal polynomial
$$(4.2.3) \qquad P_{n,q}(x) = P_{n,q}(W,x) := T_{n,q}(x)/E_{n,q}(W,I).$$

For $W \in \mathcal{F}(C^3)$ Lubinsky and Saff [84] proved the following strong asymptotic relation for $P_{n,q}(a_n z)$ with $1 < q < \infty$ and $z \in \mathbb{C} \setminus [-1,1]$:
$$(4.2.4) \qquad \frac{a_n^{1/p} P_{n,q}(a_n z)}{[\psi(z)]^n \exp(H(1/\psi(z)))} = \frac{\bigl(1 - (\psi(z))^{-2}\bigr)^{1/q} 2^{1/q-1}}{k_q}(1 + o(1)), \quad n \to \infty,$$
where H is defined at (3.0.7) and
$$k_q := \bigl[\Gamma(1/2)\Gamma((q+1)/2)/\Gamma(q/2+1)\bigr]^{1/q}.$$
Recall that $\psi(u) := u + \sqrt{u^2 - 1}$ is the Joukowski map.

Moreover, (4.2.4) holds uniformly in closed subsets of $\mathbb{C} \setminus [-1,1]$. The similar relation for Erdős-type weights on $\mathbb{R}$ was proved by Lubinsky [75]. More precise asymptotics for $W(x) = \exp(-|x|^\alpha)$, $\alpha > 0$, were established by Rakhmanov [108], Deift et al. [23] and Kriecherbauer and McLaughlin [64]. (4.2.4) also holds for $W \in \mathcal{F}(\mathrm{Lip})$ [72].

Asymptotic (4.2.4) immediately implies the following estimate:

PROPOSITION 4.2.2. *Let $W \in \mathcal{F}(C^3)$ and let $q \in (1, \infty)$. Then for any $\gamma \in (0,1)$ and for every $u \in \mathbb{C}$, satisfying*
$$1/|\psi(u)| \leq 1 - \gamma,$$

the following inequality holds:

(4.2.5) $$|P_{n,q}(a_n(u))| \geq C a_n^{-1/q} |\psi(u)|^n \big| \exp\big(H(1/\psi(u))\big)\big|$$

for all n large enough.

To prove Theorems 2.2.1 and 2.2.2, we shall need estimate (4.2.5) in a more delicate situation, namely for $W \in \mathcal{F}(\text{Lip})$ and any u from an $n^{-\varepsilon}$-neighborhood of $[-1,1]$.

PROPOSITION 4.2.3. *Let $W \in \mathcal{F}(\text{Lip})$ and let $q \in (1,\infty)$. Then there exists $\delta \in (0,1/3)$ such that for any $u \in \mathbb{C}$, satisfying*

(4.2.6) $$1/|\psi(u)| \leq 1 - Cn^{-\delta},$$

inequality (4.2.5) holds for all n large enough.

The proof is based on the existence of a weight $V_{n,q}$ that approximates $W(a_n \cdot)$ on $[-1,1]$.

LEMMA 4.2.4. *Let $W \in \mathcal{F}(\text{Lip})$. Then for each $n = 1, 2, \ldots$ and $q \in (1, \infty)$ there exists a weight $V_{n,q}$ on $[-1,1]$, satisfying the following properties:*

(4.2.7) $$0 < W(a_n x)/V_{n,q}(x) < 1, \qquad x \in [-1,1],$$

(4.2.8) $$\frac{1}{2\pi} \int_{-1}^{1} \frac{\log(V_{n,q}(x)/W(a_n x))}{\sqrt{1-x^2}} \, dx \leq Cn^{-\varepsilon_1},$$

(4.2.9) $\big|P_{n,q}(W, a_n u) a_n^{1/q} - P_{n,q}(V_{n,q}, u)\big|$
$$\leq \frac{Cn^{-\varepsilon_2} |\psi(u)|^n \big|\exp(H(W, 1/\psi(u)))\big| \, |1 - \psi^{-2}(u)|^{-1/q}}{1 - |\psi(u)|^{-1}}, \quad u \in \mathbb{C} \setminus [-1,1]$$

$$\left| \frac{P_{n,q}(V_{n,q}, u)}{2^{1/q-1} k_q^{-1} (\psi(u))^n \exp(H(V_{n,q}(a_n^{-1} \cdot), 1/\psi(u)))(1 - \psi^{-2}(u))^{-1/q}} - 1 \right|$$

(4.2.10) $$\leq |\psi(u)|^{-(1/2)n^{1/3}}, \qquad u \in \mathbb{C} \setminus [-1,1].$$

Here, ε_1 and ε_2 are positive small enough constants and C, ε_1, ε_2 are independent of n and u.

PROOF. Let $V_{n,q}$ be the weight defined by Eq. (14.38) in [72]. Then by Theorem 8.3 in [72], (4.2.7) holds and

$$\exp\left(\frac{1}{\pi} \int_{-1}^{1} \frac{\log(W(a_n x)/V_{n,q}(x))}{\sqrt{1-x^2}} \, dx \right) = 1 + O(n^{-\varepsilon_1}).$$

This implies (4.2.8).

Next, the following inequality is established in [72, Eqs. (14.35), (14.36), and (14.47)]:

(4.2.11) $\big|P_{n,q}(W, a_n u) a_n^{1/q} - P_{n,q}(V_{n,q}, u)\big|$
$$\leq \frac{\alpha_n |\psi(u)|^n \big|\exp(H(W, a)/\psi(u)))\big| \, |H_1(u)|}{1 - |\psi(u)|^{-1}}, \quad u \in \mathbb{C} \setminus [-1,1],$$

where

(4.2.12) $$\alpha_n \leq \Psi_q(A_{n,q}) + (A_{n,q} - 1).$$

Here,

(4.2.13) $$H_1(u) := \exp\left(-\frac{1}{2\pi}\int_{-\pi}^{\pi}\log(1-\cos^2 t)^{1/(2p)}\frac{1+(1/\psi(u))e^{it}}{1-(1/\psi(u))e^{it}}\,dt\right),$$

(4.2.14) $$A_{n,q} := E_{n,q}(W,I)a_n^{-(n+1/q)}/E_{n,q}(W(a_n\cdot),[-1,1]) > 1,$$

$$\Psi_q(y) := \begin{cases} 2^{1-1/q}(y^q-1)^{1/q}, & 2 \le q < \infty,\ y \ge 1 \\ 2\left(\left(\frac{1+y^2}{2}\right)^{1/(q-1)}-1\right), & 1 < q < 2,\ y \ge 1. \end{cases}$$

Then by Theorem 8.4(b) in [72],

(4.2.15) $$A_{n,q} = 1 + O(n^{-\varepsilon_3}),$$

where $\varepsilon_3 > 0$ is small enough. Since for $y \in [1,2]$,

(4.2.16) $$\Psi_q(y) \sim \begin{cases} (y-1)^{1/q}, & 2 \le q < \infty \\ y^{(q-1)/q}, & 1 < q < 2, \end{cases}$$

we obtain from (4.2.12), (4.2.14)–(4.2.16)

(4.2.17) $$\alpha_n \le Cn^{-\varepsilon_2},$$

where $\varepsilon_2 > 0$ is small enough.

Further, it is easy to compute $H_1(u)$ (see [113, Eq. (10.2.13)])

(4.2.18) $$H_1(u) = C\big(1 - \psi^{-2}(u)\big)^{-1/q}.$$

Thus (4.2.11), (4.2.17), and (4.2.18) yield (4.2.9). Finally, we note that (4.2.10) is proved in [72, p. 396]. ∎

PROOF. Proof of Proposition 4.2.3 Using (4.2.9) and (4.2.10), we obtain for every $u \in \mathbb{C}\setminus[-1,1]$

(4.2.19)
$$\begin{aligned}&|P_{n,q}(W,a_n u)|a_n^{1/q}|\psi(u)|^{-n} \\ &\ge C\big|\exp\big(H(V_{n,q}(a_n^{-1}\cdot),1/\psi(u))\big)\big|\,\big|1-\psi^{-2}(u)\big|^{-1/q}\big(1-|\psi(u)|^{-1}\big)^{-(1/2)n^{1/3}} \\ &\quad - C_1 n^{-\varepsilon_2}\big|\exp\big(H(W,1/\psi(u))\big)\big|\,\big|1-\psi^{-2}(u)\big|^{-1/q}\big(1-|\psi(u)|^{-1}\big)^{-1}.\end{aligned}$$

Next estimating the Poisson kernel, we have by (4.2.7) and (4.2.8),

$$\begin{aligned}&\big|\exp\big(H(V_{n,q}(a_n^{-1}\cdot)/W,1/\psi(u))\big)\big| \\ &\ge \exp\left(-\frac{1}{2\pi}\int_{-1}^{1}\frac{\log(V_{n,q}(x)/W(a_n x))}{\sqrt{1-x^2}}\,dx\frac{C}{1-|\psi(u)|^{-1}}\right) \\ &\ge \exp\big(-Cn^{-\varepsilon_1}(1-|\psi(u)|^{-1})^{-1}\big).\end{aligned}$$

Hence

$$\begin{aligned}&\big|\exp\big(H(V_{n,q}(a_n^{-1}\cdot),1/\psi(u))\big)\big| \\ &= \big|\exp\big(H(V_{n,q}(a_n^{-1}\cdot)/W,1/\psi(u))\big)\big|\,\big|\exp\big(H(W,1/\psi(u))\big)\big|\end{aligned}$$

(4.2.20) $$\ge \exp\big(-Cn^{-\varepsilon_1}(1-|\psi(u)|^{-1})^{-1}\big)\big|\exp\big(H(W,1/\psi(u))\big)\big|.$$

Combining (4.2.19) with (4.2.20), we have

$$(4.2.21) \quad |P_{n,q}(W, a_n u)| \geq C a_n^{-1/q} |\psi(u)|^n \big|\exp\big(H(W, 1/\psi(u))\big)\big| \Delta_n\left(\frac{1}{\psi(u)}\right),$$

where for $|w| < 1$,

$$\Delta_n(w) := |1 - w^2|^{-1/q} \bigg(\left(1 - |w|^{(1/2)n^{1/3}}\right) \exp\left(-Cn^{-\varepsilon_1}(1 - |w|)^{-1}\right)$$

$$- n^{-\varepsilon_2}(1 - |w|)^{-1} \bigg).$$

Let us choose $\delta = \theta \min(\varepsilon_1, \varepsilon_2, 1/3)$, where θ is a number from $(0, 1)$. Then there exists n_0 large enough such that

$$(4.2.22) \quad \inf_{n > n_0} \inf_{|w| \leq 1 - Cn^{-\delta}} \Delta_n(w) = C_0 > 0.$$

Thus (4.2.21) and (4.2.22) yield (4.2.5). ∎

The next result is an analogue of Proposition 4.2.3 for the orthonormal polynomials $P_{n,2}(W_1, x) = P_{n,2}(x)$ associated with the weight $W_1^2(x) = \exp(-2|x|)$. Since $W_1 \notin \mathcal{F}(\mathrm{Lip})$ (though $W_1 \in \mathcal{F}(C^3)$), we shall use a different approach based on very precise asymptotics for $P_{n,2}$ established by Kriecherbauer and McLaughlin [64]. This approach allows us to choose $\delta = 1/3 - \varepsilon$ in (4.2.6) for $W = W_1$. Note that estimates of δ are important for the limit theorems of approximation with W_1, since for this boundary weight, the constants in the corresponding relations are very sensitive with respect to δ, see Section 10.2 for more details.

PROPOSITION 4.2.5. *For any $\varepsilon \in (0, 1/3)$ and every $u \in \mathbb{C}$, satisfying*

$$(4.2.23) \quad 1/|\psi(u)| \leq 1 - Cn^{-1/3 + \varepsilon},$$

the following inequality holds:

$$(4.2.24) \quad |P_{n,2}(a_n u)| \geq C a_n^{-1/2} |\psi(u)|^n \big|\exp\big(H(1/\psi(u))\big)\big|,$$

where

$$(4.2.25) \quad a_n = \pi n / 2.$$

Note first that (4.2.25) easily follows from (1.4.12) (see also Proposition 10.1.1). The proof of Proposition 4.2.5 is based on the asymptotics for $P_{n,2}$ obtained in [64].

Since $P_{n,2}(-z) = (-1)^n P_{n,2}(z)$, $P_n(\bar{z}) = \overline{P_n(z)}$, we present the asymptotic formulae only in the first quadrant $\mathbb{C}_1$ of the complex plane. Let us define

$$C^\lambda := \{z \in \mathbb{C}_1 : |z - 1| \leq \lambda\},$$
$$B^\lambda := \{z \in \mathbb{C} : 0 \leq \mathrm{Re}\, z \leq 1,\ 0 \leq \mathrm{Im}\, z \leq \lambda\} \setminus C^\lambda,$$
$$A^\lambda := \mathbb{C}_1 \setminus (B^\lambda \cup C^\lambda).$$

Next, let

$$\psi_1(z) = \frac{1}{\pi} \int_1^{1/z} (w^2 - 1)^{-1/2}\, dw, \qquad \mathrm{Re}\, z > 0,\ z \notin [1, \infty).$$

It is known [64, Eq. (5.6)] that there exists an analytic function f_1 in a neighborhood of 1, satisfying

$$f_1(z) = \left(\frac{3}{2}\pi i \int_1^z \psi_1(y)\,dy\right)^{2/3},$$

(4.2.26) $\quad -\dfrac{2}{3}\left(n^{2/3} f_1(z)\right)^{3/2} = -n\pi i \displaystyle\int_1^z \psi_1(y)\,dy, \quad z \in \mathbb{C}_1.$

LEMMA 4.2.6. *There exists a number $\lambda_0 > 0$ such that for all $\lambda \in (0, \lambda_0]$ the following asymptotics hold:*

(i) *For $z \in A^\lambda$,*

(4.2.27) $P_{n,2}(a_n z) e^{-a_n z} = \sqrt{\dfrac{1}{4\pi a_n}} \exp\left(-n\pi i \int_1^z \psi_1(y)\,dy\right)$
$$\times \left(\frac{(z-1)^{1/4}}{(z+1)^{1/4}} + \frac{(z+1)^{1/4}}{(z-1)^{1/4}}\right) \left(1 + O\left(\frac{1}{n}\right)\right).$$

(ii) *For $z \in B^\lambda$,*

$$P_{n,2}(a_n z) e^{-a_n z} = \sqrt{\frac{2}{\pi a_n}} (1-z)^{-1/4}(1+z)^{-1/4}$$
$$\times \left[\cos\left(n\pi \int_1^z \psi_1(y)\,dy + (1/2)\arcsin z\right)(1 + O(1/\log n))\right.$$

(4.2.28) $\quad\quad + \sin\left(n\pi \displaystyle\int_1^z \psi_1(y)\,dy - (1/2)\arcsin z\right) O(1/\log n) \bigg].$

(iii) *For $z \in C^\lambda$,*

$$P_{n,2}(a_n z) e^{-a_n z} = \sqrt{\frac{1}{a_n}} \left[\frac{(z+1)^{1/4}}{(z-1)^{1/4}} \left(n^{2/3} f_1(z)\right)^{1/4}\right.$$
$$\times \text{Ai}\left(n^{2/3} f_1(z)\right)(1 + O(1/n)) - \frac{(z-1)^{1/4}}{(z+1)^{1/4}} \left(n^{2/3} f_1(z)\right)^{-1/4}$$

(4.2.29) $\quad\quad \times \text{Ai}'\left(n^{2/3} f_1(z)\right)(1 + O(1/n)) \bigg],$

where $\text{Ai}(z)$ is the Airy function [1, p. 447].

The proof of a slightly stronger form of the lemma is given in [64, Theorem 1.16]. The similar asymptotics are also valid for $P_{n,2}(W_\alpha, z)$, $1 < \alpha < \infty$.

We remark that Lemma 4.2.6(i) was established in [84, Corollary 3.3] with

(4.2.30) $\quad \gamma_n(z) := n\pi i \displaystyle\int_1^z \psi_1(y)\,dy = -H\big(W_1, 1/\psi(z)\big) - n\log|\psi(z)| + a_n z.$

It is easy to see that (4.2.30) holds for all $z \in \mathbb{C}_1 \setminus [1, \infty)$.

The following estimate of $\gamma_n(z)$ plays an essential role in the proof of Proposition 4.2.5:

LEMMA 4.2.7. *There exists a number $\lambda > 0$ such that for any $z \in B^\lambda \cup C^\lambda$, satisfying $2\pi/3 \leq \arg(z-1) \leq \pi$, the following inequality holds*

(4.2.31) $\quad\quad\quad\quad \text{Re}(\gamma_n(z)) \leq -Cn(|\psi(z)| - 1)^3.$

PROOF. Making the substitution

(4.2.32) $$z = (1/2)(w + 1/w), \quad w = \psi(z) = z + \sqrt{z^2 - 1} = Re^{i\varphi},$$

we have for some $\lambda' = \lambda'(\lambda)$ small enough

$$\begin{aligned}(4.2.33) \quad & \psi(\{z \in B^\lambda \cup C^\lambda : 2\pi/3 \leq \arg(z-1) \leq \pi\}) \\ &= \psi(\{z \in B^\lambda \cup C^\lambda : 0 \leq \operatorname{Re} z \leq 1 - |z-1|/2\}) \\ &\subseteq \left\{Re^{i\varphi} : 1 \leq R \leq 1 + \lambda', \ 0 \leq \cos\varphi \leq \frac{2}{R+1/R}\left(1 - \frac{(R-1)^2}{4R}\right)\right\}.\end{aligned}$$

Then (4.2.31) follows from (4.2.32), (4.2.33) and the following inequality:

(4.2.34) $$F(Re^{i\varphi}) \leq -(n/12)(R-1)^3,$$

where

$$R \in [1, 1+\lambda'], \quad \arccos\frac{6R - R^2 - 1}{2(R^2+1)} \leq \varphi \leq \frac{\pi}{2},$$

and λ' is small enough. Here,

$$F(Re^{i\varphi}) = F(w) := \operatorname{Re}\bigl(-H(1/w) - n\log|w| + (\pi n/4)(w + 1/w)\bigr),$$

by (4.2.30) and (4.2.25).

To prove (4.2.34), we first study the behavior of the function

$$\begin{aligned}\phi(u,y) &:= \operatorname{Re}\bigl[e^{-iy}(1 + ue^{-iy})^{-1/2}\bigr] \\ &= (1 + u^2 + 2u\cos y)^{-1/4}\cos(h(u,y)), \quad y \in [0,\pi], \ u \in [0,1],\end{aligned}$$

where

$$h(u,y) = h(y) := \frac{1}{2}\arctan\left(\frac{u\sin y}{1 + u\cos y}\right) - y.$$

A straightforward calculation shows that

(4.2.35) $$\frac{\partial\phi(u,y)}{\partial y} = (1 + u^2 + 2u\cos y)^{-5/4} G(u,y),$$

where

(4.2.36) $$G(u,y) := (1/2)u\sin y\cos(h(y)) + \bigl(1 + u^2/2 + (3/2)u\sin y\bigr)\sin(h(y)).$$

Next, we note that for $y \in [0,\pi], u \in [0,1]$,

(4.2.37) $$-\pi \leq h(y) \leq h_1(y) := \frac{1}{2}\left(\arctan\left(\frac{u\sin y}{1 + u\cos y}\right) - y\right) \leq 0.$$

Indeed, $h_1(0) = 0$, $h(\pi) = -\pi$, and

$$\begin{aligned}h'(y) &= -\frac{1 + u^2/2 + (3u\cos y)/2}{1 + u^2 + 2u\cos y} \leq -\frac{(1-u)(1-u/2)}{1 + u^2 + 2u\cos y} \leq 0, \\ h_1'(y) &= -\frac{1 + u\cos y}{2(1 + u^2 + 2u\cos y)} \leq 0.\end{aligned}$$

Hence (4.2.37) follows. Further by elementary manipulations,

(4.2.38) $$G(u,y) = u\sin(h_1(y))\cos(y/2) + \sin(h(y))(u^2/2 + u\cos y - u/2 + 1).$$

Taking into account the inequality

$$u^2/2 + u\cos y - u/2 + 1 \geq (1-u)(1-u/2) \geq 0,$$

4.2. EXTREMAL POLYNOMIALS

we obtain from (4.2.37) and (4.2.38)

(4.2.39) $$G(u, y) \leq 0, \quad y \in [0, \pi], \quad u \in [0, 1].$$

Thus by (4.2.35) and (4.2.39), $\phi(u, y)$ is decreasing in $y \in [0, \pi]$ for each $u \in [0, 1]$.
Then a straightforward calculation shows that for $t \in [0, 1)$, $R \to 1+$, and

(4.2.40) $$\frac{\pi}{2} \geq \varphi \geq \varphi_R := \arccos \frac{6R - R^2 - 1}{2(R^2 + 1)},$$

the following relations hold:

$$R^{-3}\phi(t/R^2, 2\varphi) = R^{-3} \operatorname{Re}\left[e^{-2\varphi i}(1 + (t/R^2)e^{-2\varphi i})^{-1/2}\right]$$
$$\leq R^{-3}\phi(t/R^2, \varphi_R)$$
$$= R^{-3}(1 + (t/R^2)^2 + 2(t/R^2)\cos 2\varphi_R)^{-1/4} \cos h(t/R^2, 2\varphi_R)$$
$$= (1+t)^{-1/2} - (3 + 2t)(1+t)^{-3/2}(R-1)$$
(4.2.41) $$+ (3/4)(4 + 8t + 3t^2)(1+t)^{-5/2}(R-1)^2 + O(R-1)^3,$$

where the error bound in (4.2.41) is uniform for $t \in [0, 1]$.

Next, we need the following formula for $H(z) = H(w_1, z)$ in terms of hypergeometric functions obtained in [93] (see also Lemma 10.2.4):

$$H(z) = 2n(F(-1/2, 1, 3/2, -z^2) - 1/2).$$

Hence by Euler's integral representation for hypergeometric functions $F(a, b, c, w)$ [32, Eq. (2.1(10))],

(4.2.42) $$-\frac{\partial H(1/(Re^{i\varphi}))}{\partial R} = \frac{4n}{3R^3 e^{2i\varphi}} F(1/2, 2, 5/2, -(Re^{i\varphi})^2)$$
$$= \frac{n}{R^3 e^{2\varphi i}} \int_0^1 t(1-t)^{-1/2}\left(1 + \frac{t}{R^2 e^{2\varphi i}}\right)^{-1/2} dt.$$

Then for all φ, satisfying (4.2.40), relations (4.2.41) and (4.2.42) imply

(4.2.43) $$\operatorname{Re}\left(-\frac{\partial H(1/(Re^{i\varphi}))}{\partial R}\right) = nR^{-3}\int_0^1 t(1-t)^{-1/2}\phi(t/R^2, 2\varphi)\, dt$$
$$\leq n\big(1 - (1 + \pi/2)(R-1) + (1/2 + 3\pi/4)(R-1)^2 + O((R-1)^3)\big),$$

where the error bound in (4.2.43) is independent of φ.

Further for all φ, satisfying (4.2.40),

$$\operatorname{Re}\left(-\frac{n}{R} + \frac{\pi n}{4}\left(e^{i\varphi} - \frac{1}{R^2 e^{i\varphi}}\right)\right) \leq n\left(-\frac{1}{R} + \frac{\pi(6R - R^2 - 1)(R^2 - 1)}{8(R^2+1)R^2}\right)$$
$$= n\big(-1 + (1 + \pi/2)(R-1) - (1 + 3\pi/4)(R-1)^2$$
(4.2.44) $$+ O((R-1)^3)\big).$$

Thus (4.2.43) and (4.2.44) yield

$$\frac{\partial F(Re^{i\varphi})}{\partial R} \leq n\big(-(1/2)(R-1)^2 + O(R-1)^3\big).$$

Choosing λ' small enough, we have for $1 \leq R \leq 1 + \lambda'$ and φ, satisfying (4.2.40),

(4.2.45) $$\frac{\partial F(Re^{i\varphi})}{\partial R} \leq -\frac{n}{4}(R-1)^2.$$

Since $\lim_{R \to 1+} F(Re^{i\varphi}) = 0$, we obtain (4.2.34) from (4.2.45). This proves the lemma. ∎

Proof of Proposition 4.2.5. It suffices to prove (4.2.24) in the first quadrant $\mathbb{C}_1$ of the complex plane. Since $\mathbb{C}_1 = A^\lambda \cup B^\lambda \cup C^\lambda$, we consider the following cases:

Case 1. Let $u \in A^\lambda$. Then $1/|\psi(u)| \leq C < 1$ (i.e. (4.2.23) is satisfied) and

$$(4.2.46) \qquad \left| \frac{(u-1)^{1/4}}{(u+1)^{1/4}} + \frac{(u+1)^{1/4}}{(u-1)^{1/4}} \right| = 2|1 - 1/\psi^2(u)|^{-1/2} \geq \sqrt{2}.$$

Thus (4.2.24) follows from (4.2.27), (4.2.30), and (4.2.46).

Case 2. Let $u \in B^\lambda$, $\pi/2 < \arg(u-1) < 2\pi/3$. Then $u \in A^{\sqrt{3}\lambda/2}$ (i.e. (4.2.23) is satisfied), and (4.2.24) follows from (4.2.27), (4.2.30), and (4.2.46).

Case 3. Let $u \in B^\lambda$, $2\pi/3 \leq \arg(u-1) \leq \pi$. Then

$$(4.2.47) \qquad \min_{z \in B^\lambda} |(1-z^2)^{-1/4}| = C > 0.$$

Next for all n large enough,

$$\left| \cos\left(\pi n \int_1^u \psi_1(y)\,dy + \frac{1}{2}\arcsin u\right)\left(1 + O\left(\frac{1}{\log n}\right)\right) \right.$$
$$\left. + \sin\left(\pi n \int_1^u \psi_1(y)\,dy - \frac{1}{2}\arcsin u\right) O\left(\frac{1}{\log n}\right) \right|$$
$$\geq \frac{1}{2} \exp\left[\operatorname{Re}\left(-n\pi i \int_1^u \psi_1(y)\,dy\right)\right]\left[\exp\left(\frac{1}{2}\min_{z \in B^\lambda} \operatorname{Re}(-i\arcsin z)\right)\right.$$
$$\times \left(1 + O\left(\frac{1}{\log n}\right)\right) - \exp\left(\frac{1}{2}\max_{z \in B^\lambda} \operatorname{Re}(i\arcsin z)\right) O\left(\frac{1}{\log n}\right)$$
$$- \exp\left(\operatorname{Re}\left(2n\pi i \int_1^u \psi_1(y)\,dy\right)\right) \max_{z \in B^\lambda}\left(\left|\exp\left(\frac{i}{2}\arcsin z\right)\right.\right.$$
$$\left.\left.\times \left(1 + O\left(\frac{1}{\log n}\right)\right)\right| + \left|\exp\left(-\frac{i}{2}\arcsin z\right) O\left(\frac{1}{\log n}\right)\right|\right)\right]$$
$$\geq \frac{1}{2} \exp\left(\operatorname{Re}\left(-n\pi i \int_1^u \psi_1(y)\,dy\right)\right)\left(C_1 - C_2 \exp\left(\operatorname{Re}\left(2n\pi i\right.\right.\right.$$
$$(4.2.48) \qquad \left.\left.\left.\times \int_1^u \psi_1(y)\,dy\right)\right)\right).$$

Further by Lemma 4.2.7, we have that there exists $\lambda > 0$ such that for $u \in B^\lambda$, $2\pi/3 \leq \arg(u-1) \leq \pi$,

$$\operatorname{Re}\left(2n\pi i \int_1^u \psi_1(y)\,dy\right) \leq -Cn(|\psi(u)| - 1)^3.$$

Hence if, in addition, u satisfies (4.2.23), then for all n large enough we obtain

$$(4.2.49) \qquad \exp\left(\operatorname{Re}\left(2n\pi i \int_1^u \psi_1(y)\,dy\right)\right) \leq \exp(-Cn^{3\varepsilon}).$$

4.2. EXTREMAL POLYNOMIALS

Then (4.2.28), (4.2.30), and (4.2.47)–(4.2.49) yield (4.2.24).

Case 4. Let $u \in C^\lambda$. We first note that for the analytic function $f_1(u)$ defined by (4.2.26),

$$(4.2.50) \qquad f_1(1) = 0, \qquad f_1'(1) = 2^{1/3}$$

(see [64, p. 316]). Then (4.2.50) implies the inequality

$$(4.2.51) \qquad n^{2/3}|f_1(u)| \geq Cn^{2\varepsilon} \quad \text{for} \quad Cn^{-2/3+2\varepsilon} \leq |1-u| \leq \lambda,$$

if λ is small enough and n large enough. Next note that (4.2.51) holds for any $u \in C^\lambda$, satisfying (4.2.23), since

$$|1-u| \geq \frac{|1-u^2|}{2+\lambda} = \frac{1}{4(2+\lambda)} \left| \frac{\psi^2(u)-1}{\psi(u)} \right| \geq Cn^{-2/3+2\varepsilon}.$$

If λ is small enough, then (4.2.50) also implies

$$(4.2.52) \qquad |\arg f_1(u) - \arg(u-1)| \leq \pi/12, \qquad u \in C^\lambda.$$

Further, we need the following asymptotics for the Airy function $Ai(z)$ and its derivative (see [1, pp. 448–449]):

$$(4.2.53) \; Ai(z) = \frac{1}{\pi}\sqrt{\frac{2}{3}} K_{1/3}\left(\frac{2}{3}z^{3/2}\right)$$
$$= \frac{1}{2}\pi^{-1/2}z^{-1/4}\exp\left(-\frac{2}{3}z^{3/2}\right)(1+O(|z|^{-3/2})), \quad -\pi < \arg z < \pi;$$

$$(4.2.54) \qquad Ai'(z) = \frac{1}{\pi}\left(\frac{z}{\sqrt{3}}\right) K_{2/3}\left(\frac{2}{3}z^{3/2}\right)$$
$$= -\frac{1}{2}\pi^{-1/2}z^{1/4}\exp\left(-\frac{2}{3}z^{3/2}\right)(1+O(|z|^{-3/2})), \quad -\pi < \arg z < \pi;$$

$$(4.2.55) \qquad Ai(-z) = \pi^{-1/2}z^{-1/4}\left(\sin\left(\frac{2}{3}z^{3/2}+\pi/4\right)(1+O(|z|^{-3}))\right.$$
$$\left. - \cos\left(\frac{2}{3}z^{3/2}+\pi/4\right)O(|z|^{-3/2})\right), \quad -2\pi/3 < \arg z < 2\pi/3;$$

$$(4.2.56) \qquad Ai'(-z) = -\pi^{-1/2}z^{1/4}\left(\cos\left(\frac{2}{3}z^{3/2}+\pi/4\right)(1+O(|z|^{-3}))\right.$$
$$\left. - \sin\left(\frac{2}{3}z^{3/2}+\pi/4\right)O(|z|^{-3/2})\right), \quad -2\pi/3 < \arg z < 2\pi/3.$$

Here $K_\nu(z)$ is Macdonald's function [122]. Since Ai is an entire function, we can replace z wth $e^{i\pi}z$ in (4.2.55) and (4.2.56) and obtain the formulae

$$Ai(z) = \frac{1}{2}\pi^{-1/2}z^{-1/4}\left(\exp\left(-\frac{2}{3}z^{3/2}\right) - i\exp\left(\frac{2}{3}z^{3/2}\right)\right)$$

(4.2.57)
$$\times \left(1 + O(|z|^{-3/2})\right), \quad \pi/3 < \arg z < 5\pi/3;$$

$$Ai'(z) = -\frac{1}{2}\pi^{-1/2}z^{1/4}\left(\exp\left(-\frac{2}{3}z^{3/2}\right) + i\exp\left(\frac{2}{3}z^{3/2}\right)\right)$$

(4.2.58)
$$\times \left(1 + O(|z|^{-3/2})\right), \quad \pi/3 < \arg z < 5\pi/3.$$

Note that all error bounds in (4.2.53)–(4.2.58) are independent of $\arg z$, if $\arg z$ belongs to closed subsets of the corresponding intervals.

Now we consider the following two cases:

Case 4(a). Let $u \in C^\lambda$ and $0 < \arg(u-1) \le 3\pi/4$. Then by (4.2.52), we have for $\lambda > 0$ small enough

$$-\pi/12 \le \arg f_1(u) \le 5\pi/6.$$

Next using (4.2.26), (4.2.29), (4.2.51), (4.2.53), and (4.2.54), we get

$$\left|P_{n,2}(a_n u)e^{-a_n u}\right| = \frac{1}{2}\pi^{-1/2}\sqrt{\frac{1}{a_n}}\left|\frac{(u+1)^{1/4}}{(u-1)^{1/4}}(1 + O(n^{-3\varepsilon}))\right.$$

$$+ \frac{(u-1)^{1/4}}{(u+1)^{1/4}}(1 + O(n^{-3\varepsilon}))\left|\left|\exp\left(-\frac{2}{3}(n^{2/3}f_1(u))^{3/2}\right)\right|\right.$$

(4.2.59)
$$\ge Ca_n^{-1/2}\exp\left(\operatorname{Re}\left(-n\pi i\int_1^u \psi_1(y)\,dy\right)\right).$$

Finally, (4.2.59) and (4.2.30) yield (4.2.24).

Case 4(b). Let $u \in C^\lambda$ and $3\pi/4 < \arg(u-1) \le \pi$. Then by (4.2.52), we have for $\lambda > 0$ small enough

$$-2\pi/3 \le \arg f_1(u) \le 13\pi/12.$$

Next using (4.2.29), (4.2.51), (4.2.57), and (4.2.58), we get

$$\left|P_{n,2}(a_n u)e^{-a_n u}\right|$$

$$= \frac{1}{2}\pi^{-1/2}\sqrt{\frac{1}{a_n}}\left|\left(\frac{(u+1)^{1/4}}{(u-1)^{1/4}}\left(\exp\left(-\frac{2}{3}(n^{2/3}f_1(u))^{3/2}\right)\right.\right.\right.$$

$$\left.\left.- i\exp\left(\frac{2}{3}(n^{2/3}f_1(u))^{3/2}\right)\right)(1 + O(n^{-3\varepsilon}))$$

$$+ \frac{(u-1)^{1/4}}{(u+1)^{1/4}}\left(\exp\left(-\frac{2}{3}(n^{2/3}f_1(u))^{3/2}\right)\right.$$

(4.2.60)
$$\left.\left.+ i\exp\left(\frac{2}{3}(n^{2/3}f_1(u))^{3/2}\right)\right) \times (1 + O(n^{-3\varepsilon}))\right|.$$

Further by (4.2.23), (4.2.26), and Lemma 4.2.7,

$$Re\left(\frac{4}{3}(n^{2/3}f_1(u))^{3/2}\right) = Re\left(2\pi ni\int_1^u \psi_1(u)\,du\right)$$
(4.2.61)
$$\leq -Cn(|\psi(z)|-1)^3 \leq -Cn^{3\varepsilon} < 0.$$

Therefore (4.2.60) and (4.2.61) imply

$$|P_{n,2}(a_nu)e^{-a_nu}| \geq Ca_n^{-1/2}\left|\frac{(u+1)^{1/4}}{(u-1)^{1/4}}(1+O(n^{-3\varepsilon}))\right.$$

$$\left. + \frac{(u-1)^{1/4}}{(u+1)^{1/4}}(1+O(n^{-3\varepsilon}))\right|\exp\left(Re\left(-n\pi i\int_1^u \psi_1(y)\,dy\right)\right)$$

(4.2.62)
$$\geq Ca_n^{-1/2}\exp\left(Re\left(-n\pi i\int_1^u \psi_1(y)\,dy\right)\right).$$

Finally, (4.2.62) and (4.2.30) yield (4.2.4). This completes the proof of Proposition 4.2.5. ∎

4.3. Polynomial Inequalities in the Complex Plane

Here we consider some weighted polynomial inequalities in the complex plane and in the real axis. In particular, we discuss two equivalent forms of the estimate for $|P(z)|$, where $P \in \mathcal{P}_n$ and $z \in \mathbb{C}$. The first of them for $I = \mathbb{R}$ and a convex W was established by Mhaskar and Saff [94] (see also [84]). The following extension of this inequality to weights on I was given in [72, p. 260]:

PROPOSITION 4.3.1. *Let $W \in \mathcal{F}_C$. Then for any $P \in \mathcal{P}_n$ and every $z \in \mathbb{C} \setminus [a_n, a_n]$, the following inequalities hold:*

(4.3.1) $\quad |P(z)| \leq \exp(-V^{\mu_n}(z) + c_n)\|P\|_{L_{\infty,W}[-a_n, a_n]},$

(4.3.2) $\quad |P(z)|W(|z|) \leq \exp(U_n(z))\|P\|_{L_{\infty,W}[-a_n, a_n]},$

where

$$V^{\mu_t}(z) := \int_{-a_t}^{a_t} \log\left|\frac{1}{z-s}\right|\sigma_t(s)\,ds$$

is the equilibrium potential with the density σ_t of the equilibrium measure μ_t defined by

$$\sigma_t(s) = \frac{1}{\pi^2}\sqrt{a_t^2 - s^2}\int_{-a_t}^{a_t}\frac{Q'(x) - Q'(s)}{(x-s)\sqrt{a_t^2 - x^2}}\,dx,$$

$$c_t := \int_0^t \log(2/a_s)\,ds$$

is a constant and

$$U_n(z) := -V^{\mu_n}(z) - Q(|z|) + c_n.$$

Note that (4.3.2) is a direct consequence of (4.3.1).

Certain technical estimates of $U_n(z)$ are given in the following proposition:

PROPOSITION 4.3.2. *Let $W \in \mathcal{F}$. Then the following statements hold:*

(a) *For any $K > 1$ there exists C and n_0 such that for $n > n_0$ and $x \in [a_n, a_{Kn}]$,*

(4.3.3) $$U_n(x) \leq -Cn(x/a_n - 1)^{3/2}.$$

(b) *For any $K > 1$ there exists C, n_0, and $\varepsilon > 0$ such that for $n > n_0$ and $x \in (a_{Kn}, c)$*

(4.3.4) $$U_n(x) \leq -Cn^\varepsilon.$$

(c) U_n *is decreasing and negative in* (a_n, c).

(d) *If $r \geq 2n$, then*
$$U_n(a_r) \leq -Cr^\varepsilon,$$
where C and ε are independent of n and r.

(e) *If $r \geq 2n$ and $a_r \leq x \leq a_{r+1}$, then*
$$U_n(x) \leq -Cr^\varepsilon,$$
where C and ε are independent of n, r, and x.

Statements (a)–(d) of the proposition are established in [72, p. 101], while (e) easily follows from (c) and (d).

Next, we establish the inequality that is equivalent to (4.3.1) but more suitable for our purposes. For $Q = |x|^\alpha$, $\alpha > 0$, $I = \mathbb{R}$, the corresponding estimate was given in [93]. For $W \in \mathcal{F}_C$, it is possible to deduce this inequality from (4.3.1) (cf. [93]) but the direct proof is shorter.

PROPOSITION 4.3.3. *Let $W \in \mathcal{F}_M$. Then for any $P \in \mathcal{P}_n$ and any $z \in \mathbb{C} \setminus [-a_n, a_n]$,*

(4.3.5) $$|P(z)| \leq \exp\bigl(\operatorname{Re}(H(1/\psi(z/a_n))) + n\log|\psi(z/a_n)|\bigr)\|P\|_{L_\infty,W[-a_n,a_n]}.$$

PROOF. We may assume without loss of generality that $\|P\|_{L_\infty,W[-a_n,a_n]} = 1$, and this is equivalent to the inequality

(4.3.6) $$F_1(t) := \log|P(a_n \cos t)| - Q(a_n \cos t) \leq 0, \qquad t \in [-\pi, \pi).$$

Next by the substitution $z = (a_n/2)(w + w^{-1})$, $0 < |w| \leq 1$, i.e., $w = 1/\psi(z/a_n)$, we transform (4.3.5) to the equivalent form

(4.3.7) $$F(w) := \log\bigl|P((a_n/2)(w + w^{-1}))\bigr| - \operatorname{Re} H(w) + n\log(w) \leq 0,$$

where $0 < |w| < 1$. Note that F is a subharmonic function in the domain $S = \{w \in \mathbb{C}: 0 < |w| < 1\}$.

Further, we estimate F on the boundary of S. Using Proposition 3.1.1(c) and taking account of continuity of $\log|P_n((a_n/2)(w + w^{-1}))| + n\log|w|$ for $|w| > 0$, we have from (4.3.6)

(4.3.8) $$\lim_{w \to e^{it}} F(w) = F_1(t) \leq 0, \qquad t \in [-\pi, \pi).$$

Furthermore, for the exceptional point $w = 0$, we get

(4.3.9) $$\lim_{w \to 0}(-\operatorname{Re} H(w)) = -(1/(2\pi))\int_{-\pi}^{\pi} Q(a_n \cos t)\,dt,$$

$$\lim_{w \to 0}\bigl(\log|P((a_n/2)(w + w^{-1}))| + n\log|w|\bigr)$$

(4.3.10) $$= \lim_{w \to 0} \log|w^n P((a_n/2)(w + w^{-1}))| = \log\bigl(|c_0|/(a_n/2)^n\bigr),$$

where c_0 is the leading coefficient of P.

Then (4.3.9) and (4.3.10) imply

(4.3.11) $$\lim_{w \to 0} F(w) < +\infty.$$

Taking account of (4.3.8), (4.3.11) and applying the maximum modulus principle for subharmonic functions [86] to F, we obtain (4.3.7). Hence (4.3.5) follows. ∎

Remarks

(a) Since the maximum modulus principle implies that

(4.3.12) $$\lim_{w \to w_0} F(w) \leq 0$$

for any boundary point w_0 including the exceptional point $w_0 = 0$ (see [86]), we have from (4.3.9), (4.3.10), and (4.3.12) the following weighted analogue of the Chebyshev estimate for the leading coefficient of a polynomial [114, p. 69]: if $W \in \mathcal{F}_M$, then for any polynomial $P(z) = \sum_{k=0}^{n} c_k x^{n-k}$,

(4.3.13) $$|c_0| \leq (2/a_n)^n \exp\left((1/(2\pi)) \int_{-\pi}^{\pi} Q(a_n \cos t)\,dt\right) \|P\|_{L_\infty, W[-a_n, a_n]}.$$

(b) Inequalities (4.3.1) and (4.3.5) are the weighted analogues of the Bernstein inequality
$$|P(z)| \leq |\psi(z)|^n \|P\|_{L_\infty[-1,1]}, \qquad z \in \mathbb{C} \setminus [-1, 1].$$

4.4. Proofs of Theorems 2.3.1 and 2.3.2

Proof of Theorem 2.3.1. Inequality (2.3.3) is trivial for $z \in [-a_n, a_n]$. If $z \in \mathbb{C} \setminus [-a_n, a_n]$, then using Theorem 3.4.1 and Proposition 4.3.3, we have

$$\begin{aligned}|P(z)| &\leq \exp(\operatorname{Re}(H(1/\psi(z/a_n)) + n \log|\psi(z/a_n)|)\|P\|_{L_\infty, W[-a_n, a_n]} \\ &\leq \exp(b_n|z|)\|P\|_{L_\infty, W[-a_n, a_n]}.\end{aligned}$$

Hence (2.3.3) follows. ∎

Proof of Theorem 2.3.2. We first prove the theorem for $p = \infty$. Inequality (2.3.4) is trivial for $k = 0$. Next using the Caushy integral formula for derivatives and applying Theorem 2.3.1, we obtain for any $r > 0$ and $1 \leq k \leq n$

$$\begin{aligned}|c_k| &= \frac{1}{2\pi}\left|\int_{|z|=r} \frac{P(z)}{z^{k+1}}\,dz\right| \leq r^{-k} \max_{|z|=r} |P(z)| \\ &\leq e^{b_n r - k \log r} \|P\|_{L_\infty, W(I)}.\end{aligned}$$ (4.4.1)

Choosing $r = k/b_n$, we have from (4.4.1)

(4.4.2) $$|c_k| \leq \frac{b_n^k}{(k/e)^k} \|P\|_{L_\infty, W(I)}.$$

Then using Stirling's formula
$$k! = (k/e)^k \sqrt{2\pi k} \exp(\tau/(12k)), \qquad \tau \in (0,1),$$
we obtain from (4.4.2) for $1 \leq k \leq n$

(4.4.3) $$|c_k| \leq \sqrt{2\pi} \, e^{1/12} \frac{\sqrt{k}\, b_n^k}{k!} \|P\|_{L_\infty, W(I)} < \frac{3\sqrt{k}\, b_n^k}{k!} \|P\|_{L_\infty, W(I)}.$$

Thus (2.3.4) follows from (1.4.27) and (4.4.3).

4. POLYNOMIAL INEQUALITIES WITH EXPONENTIAL WEIGHTS

To prove statement (b), we use (2.3.4) for $P(x)$ replaced with

$$P((1-\varepsilon)x) = \sum_{k=0}^{n} c_k (1-\varepsilon)^k x^k$$

and take account of Nikolskii-type inequality (4.1.10). Then

$$|c_k|(1-\varepsilon)^k \leq \frac{3\sqrt{k}\, b_n^k}{k!} \max_{x \in I} |P((1-\varepsilon)x)| W(x)$$

(4.4.4)
$$\leq \frac{C\sqrt{k}\, b_n^k}{k!} (n/a_n)^{1/p} \|P\|_{L_{p,W}(I)}.$$

Next, (3.2.6) and (4.4.4) yield

(4.4.5)
$$|c_k| \leq \frac{C\sqrt{k}\, b_n^{k+1/p}}{k!(1-\varepsilon)^k} \|P\|_{L_{p,W}(I)}.$$

Finally, (2.3.5) follows from (4.4.5) and (1.4.28). ∎

CHAPTER 5

Entire Functions of Exponential Type and their Approximation Properties

In this chapter, we consider certain properties of entire functions of exponential type which are used in proofs of the limit theorems. Most of these properties in more general settings are established in a number of papers and books. Here, for the convenience of the reader, we give a direct proof of each property.

5.1. Entire Functions of Exponential Type

We recall that B_σ with $\sigma > 0$ is the class of all entire functions g of exponential type σ, i.e., satisfying the condition: for every $\varepsilon > 0$ there exists $C = C(g, \varepsilon)$ such that

(5.1.1) $$|g(z)| \leq C e^{\sigma(1+\varepsilon)|z|}, \qquad z \in \mathbb{C}.$$

It is easy to see that if $g \in B_\sigma$, then $g(\tau \cdot) \in B_{\sigma\tau}$, $\sigma > 0$, $\tau > 0$.

Let $M_{A,N}$ be a set of all measurable function g on $\mathbb{R}$, satisfying the condition

(5.1.2) $$|g(x)| \leq A(1+|x|)^N$$

for all $x \in \mathbb{R}$, where $N \geq 0$ is an integer.

We first establish some inequalities for entire functions of exponential type.

PROPOSITION 5.1.1. (a) If $g \in B_\sigma \cap M_{A,N}$, then

(5.1.3) $$|g(z)| \leq 2^{N/2} A (1+|z|)^N e^{\sigma |\text{Im } z|}.$$

(b) If $g \in B_\sigma$ satisfies (5.1.2) for all $x \in E$, where $|\mathbb{R} \setminus E| \leq b < \infty$, then $g \in M_{CA,N}$, where C depends only on b and σ.

(c) If $g \in B_\sigma \in L_p(\mathbb{R})$, $0 < p \leq \infty$, then

(5.1.4) $$\|g\|_{L_\infty(\mathbb{R})} \leq C \sigma^{1/p} \|g\|_{L_p(\mathbb{R})},$$

where C is an absolute constant.

PROOF. (a) The function $G(z) := g(z) e^{\sigma(1+\varepsilon)iz}(z+i)^{-N}$ is analytic in the upper-half-plane H_+ and by (5.1.1), (5.1.2),

$$\sup_{x \in \mathbb{R}} |G(x)| \leq \sup_{x \in \mathbb{R}} |g(x)|(1+x^2)^{-N/2} \leq 2^{N/2} A; \quad \sup_{y>0} |G(iy)| \leq C(g, \varepsilon).$$

Next applying the Phragmen-Lindelöf principle [67] to two quadrants of H_+, we see that G is uniformly bounded on H_+. Then using the Phragmen-Lindelöf theorem [67] to G on H_+, we conclude that $\sup_{z \in H_+} |G(z)| \leq 2^{N/2} A$, i.e.,

$$|g(x+iy)| \leq 2^{N/2} A (1+|z|)^N e^{\sigma(1+\varepsilon)|y|}, \qquad y \geq 0.$$

Letting $\varepsilon \to 0$, we obtain (5.1.3) for $\text{Im } z \geq 0$. The similar argument shows that (5.1.3) holds for $\text{Im } z \leq 0$ as well.

(b) To prove this statement, we use the following result announced by Levin [68] and proved independently by Logvinenko, Sereda [73] and Kacnel'son [57]: Let a set $E \subseteq \mathbb{R}$ satisfy the (L, ε)-condition, i.e., there exist $L > 0$ and $\varepsilon > 0$ such that $\inf_{y \in \mathbb{R}} |E \cap [y, y+L]| \geq \varepsilon$; then for $g \in B_\sigma \cap L_\infty(E)$,

$$(5.1.5) \qquad \|g\|_{L_\infty(\mathbb{R})} \leq C(L, \varepsilon, \sigma) \|g\|_{L_\infty(E)}.$$

It is easy to see that if $0 < |\mathbb{R} \setminus E| \leq b < \infty$, then E satisfies the $(2b, b)$-condition. Next, the function

$$g_1(z) := g(z)(\sin z/z)^N$$

belongs to $B_{N+\sigma}$ and $\|g_1\|_{L_\infty(E)} \leq C_1 A$. Hence by (5.1.5), $\|g_1\|_{L_\infty(\mathbb{R})} \leq C_2 A$. So for every

$$x \in \Omega := \bigcup_{|\ell|=0}^{\infty} \{x \in \mathbb{R} : (4\ell+1)\pi/4 \leq x \leq (4\ell+3)\pi/4\}$$

the inequality

$$|g(x)| \leq C_3 A (1 + |x|)^N$$

holds true.

Further, the function $g_2(z) := g(z + \pi/2) \in B_\sigma$ and satisfies the inequality

$$|g_2(x)| \leq A(1 + |x + \pi/2|)^N \leq C_4 A(1 + |x|)^N, \qquad x \in E + \pi/2.$$

Then replacing g with g_2 in the previous estimates, we get

$$|g_2(x)| \leq C_5 A (1 + |x|)^N, \qquad x \in \Omega.$$

Hence

$$|g(x)| \leq C_6 A (1 + |x|)^N, \qquad x \in \Omega + \pi/2.$$

Since $\overline{\mathbb{R} \setminus \Omega} = \Omega + \pi/2$, we obtain $g \in M_{CA, N}$.

(c) We first note that the function $g_\sigma(z) := g(z/\sigma)$ belongs to $B_1 \cap L_p(\mathbb{R})$ and

$$|E_1| := \left|\left\{x \in \mathbb{R} : |g_\sigma(x)| > \|g_\sigma\|_{L_p(\mathbb{R})}\right\}\right| \leq 1, \qquad 0 < p < \infty.$$

Then the set $E := \mathbb{R} \setminus E_1$ satisfies the condition $|\mathbb{R} \setminus E| \leq 1$ and $\|g_\sigma\|_{L_\infty(E)} \leq \|g_\sigma\|_{L_p(\mathbb{R})}$. Applying statement (b) to g_σ for $N = 0$, $A = \|g_\sigma\|_{L_p(\mathbb{R})}$, $b = 1$, we have

$$\|g\|_{L_\infty(\mathbb{R})} = \|g_\sigma\|_{L_\infty(\mathbb{R})} \leq C \|g_\sigma\|_{L_p(\mathbb{R})} = C \sigma^{1/p} \|g\|_{L_p(\mathbb{R})},$$

where C is an absolute constant. Thus (5.1.4) follows. ∎

Next, we consider the compactness properties of functions from B_σ.

PROPOSITION 5.1.2. *Let*

$$g_n(z) = \sum_{k=0}^{\infty} c_{k,n} z^k, \qquad n = 1, 2, \ldots$$

be a sequence of entire functions of exponential type σ. Then the following statements hold:

(a) *If for some integer $N \geq 0$,*

(5.1.6) $$|c_{k,n}| \leq \frac{C(k+1)^N \sigma^k}{k!}, \qquad k = 0, 1, \ldots, \quad n = 1, 2, \ldots,$$

where C is independent of k and n, then there exist a subsequence $\{g_{n_s}\}_{s=1}^{\infty}$ and a function $g_0 \in B_\sigma$ such that g_{n_s} converges uniformly to g_0 as $s \to \infty$ in each bounded subset of $\mathbb{C}$.

(b) *If*

$$\sup_n \|g_n\|_{L_p(\mathbb{R})} < \infty, \qquad 0 < p \leq \infty,$$

then there exist a subsequence $\{g_{n_s}\}_{s=1}^{\infty}$ and $g_0 \in B_\sigma$ such that g_{n_s} converges uniformly to g_0 as $s \to \infty$ in each bounded subset of $\mathbb{C}$.

(c) *If $\sigma \geq 1$ and $g_n \in M_{A,N}$, $n = 1, 2, \ldots$, where A and N are independent of n, then there exist a subsequence $\{g_{n_s}\}_{s=1}^{\infty}$ and $g_0 \in B_\sigma$ such that g_{n_s} converges uniformly to g_0 as $s \to \infty$ in each bounded subset of $\mathbb{C}$.*

PROOF. (a) By (5.1.6), the coefficients $c_{k,n}$ are uniformly bounded for each k. So using the Cantor diagonal process, it is possible to select a subsequence of functions $\{g_{n_s}\}_{s=1}^{\infty}$ such that

(5.1.7) $$\lim_{s \to \infty} c_{n_s,k} = c_k, \qquad k = 0, 1, \ldots.$$

Then setting

$$g_0(z) := \sum_{k=0}^{\infty} c_k z^k$$

and taking account of the inequality

(5.1.8) $$|c_k| \leq \frac{C(k+1)^N \sigma^k}{k!}$$

that follows from (5.1.6) and (5.1.7), we obtain

$$|g_0(z)| \leq C \sum_{k=0}^{\infty} \frac{(k+1)^N \sigma^k |z|^k}{k!} \leq P_N(\sigma|z|) e^{\sigma|z|},$$

where $P_N \in \mathcal{P}_N$. Hence $g_0 \in B_\sigma$.

Next for every $z \in \mathbb{C}$ with $|z| \leq M$ and for any natural L, we have

(5.1.9) $$|g_0(z) - g_n(z)| \leq \sum_{k=0}^{L} |c_k - c_{k,n}| M^k + \sum_{k=L+1}^{\infty} (|c_k| + |c_{k,n}|) M^k.$$

It follows from (5.1.6)–(5.1.8) that for every $\varepsilon > 0$ and any $M > 0$ there exist $L = L(\varepsilon, M)$ and $n_0 = n_0(\varepsilon, L, M)$ such that for $n > n_0$, the inequalities

$$\sum_{k=L+1}^{\infty} (|c_k| + |c_{k,n}|) M^k < \varepsilon/2, \qquad \sum_{k=0}^{L} |c_k - c_{k,n}| M^k < \varepsilon/2$$

hold. Hence statement (a) follows from (5.1.9).

(b) Note first that the following Bernstein inequality for entire functions of exponential type holds [114, p. 208]:

(5.1.10) $$\|g^{(k)}\|_{L_\infty(\mathbb{R})} \leq \sigma^k \|g\|_{L_\infty(\mathbb{R})}, \qquad g \in B_\sigma \cap L_\infty(\mathbb{R}).$$

Then combining (5.1.9) with (5.1.10), we have that (5.1.6) holds with $N = 0$. Thus statement (b) follows from (a).

(c) To prove this statement, we use the following generalization of (5.1.10) obtained by Bernstein [12] (see also [10]): if $g \in B_\sigma$ satisfies the estimate
$$|g(x)| \le A(1+x^2)^{N/2}, \qquad x \in \mathbb{R}, \quad N \in \mathbb{N},$$
then
$$|g^{(k)}(x)| \le A\left|\left[(x+i)^N e^{-i\sigma x}\right]^{(k)}\right|.$$
In particular, for $\sigma \ge 1$,
$$(5.1.11) \qquad \frac{|g^{(k)}(0)|}{k!} \le A \sum_{s=0}^{\min(k,N)} \frac{\binom{N}{s}\sigma^{k-s}}{(k-s)!} \le A\sigma^k(1+1/\sigma)^N \le A 2^N \sigma^k.$$
Then statement (c) follows from (5.1.11) and statement (a). ∎

The following simple estimates immediately follow from Proposition 5.1.1(a) and inequality (5.1.1):

PROPOSITION 5.1.3. *Let $R > 1$. Then the following statements hold:*
(a) *If $g \in B_\sigma \cap M_{A,N}$, then*
$$(5.1.12) \qquad \max_{|w|=R}\left|g\left(\frac{a_n}{2}\left(w+\frac{1}{w}\right)\right)\right| \le CAa_n^N R^N \exp\left(\frac{a_n\sigma}{2}\left(R-\frac{1}{R}\right)\right),$$
where $C = C(N)$.
(b) *If $g \in B_\sigma$, then for every $\varepsilon > 0$ there exists $C = C(g,\varepsilon)$ such that for all $R > 1$,*
$$(5.1.13) \qquad \max_{|w|=R}\left|g\left(\frac{a_n}{2}\left(w+\frac{1}{w}\right)\right)\right| \le C\exp\left(\frac{a_n\sigma(1+\varepsilon)}{2}\left(R+\frac{1}{R}\right)\right).$$

Remarks

(a) Proposition 5.1.1(a) for $N = 0$ is well known [67, 114], while for $N > 0$ it was proved by Genchev [53]. Proposition 5.1.1(b) in a more general situation was established by the author [45, 47]. The first proof of Proposition 5.1.1(c) was apparently given by Plancherel and Polya [102]. Nikolskii [100, 101] independently proved it for $1 \le p \le \infty$. Various generalizations of (5.1.4) can be found in [47, 48, 101, 102]. Here we give a new proof of Proposition 5.1.1(c).

(b) Various versions of Proposition 5.1.2(a) for $N = 0$ and Proposition 5.1.2(b) for $p \ge 1$ are discussed in [4, 12, 101, 114]. Proposition 5.1.2(c) was established in [12].

5.2. Approximation Properties of Entire Functions of Exponential Type

We first establish the existence of best entire approximants.

PROPOSITION 5.2.1. *If $A_\sigma(f, L_p) < \infty$, $0 < p \le \infty$, then there exists $g_\sigma \in B_\sigma$ such that*
$$(5.2.1) \qquad A_\sigma(f, L_p) = \|f - g_\sigma\|_{L_p(\mathbb{R})}.$$

5.2. APPROXIMATION PROPERTIES OF ENTIRE FUNCTIONS OF EXPONENTIAL TYPE

PROOF. Without loss of generality we assume that $f \in L_p(\mathbb{R})$ and $f \notin B_\sigma$. Then there exists a sequence $\{g_{\sigma,n}\}_{n=1}^\infty$ such that $g_{\sigma,n} \in B_\sigma$ and

(5.2.2) $\qquad \|f - g_{\sigma,n}\|_{L_p(\mathbb{R})} < A_\sigma(f, L_p) + \|f\|_{L_p(\mathbb{R})}/n, \qquad n = 1, 2, \ldots.$

Next by (1.3.1), (1.3.4), and (5.2.2),

$$\|g_{\sigma,n}\|_{L_p(\mathbb{R})} \le 2^{1/\mu(p)-1}\bigl(\|f - g_{\sigma,n}\|_{L_p(\mathbb{R})} + \|f\|_{L_p(\mathbb{R})}\bigr) \le 2^{1/\mu(p)}\|f\|_{L_p(\mathbb{R})}.$$

Then by Proposition 5.1.2(b), there exist a subsequence $\{g_{\sigma,n_s}\}_{s=1}^\infty$ and $g_\sigma \in B_\sigma$ such that for every $M > 0$

(5.2.3) $\qquad \lim_{s\to\infty} \|g_\sigma - g_{\sigma,n_s}\|_{L_p[-M,M]} = 0.$

Further using (1.3.1), (5.2.2), and (5.2.3), we have

$$\|f - g_\sigma\|_{L_p(\mathbb{R})}^{\mu(p)} \ge A_\sigma^{\mu(p)}(f, L_p) = \lim_{s\to\infty} \|f - g_{\sigma,n_s}\|_{L_p(\mathbb{R})}^{\mu(p)}$$
$$\ge \|f - g_\sigma\|_{L_p[-M,M]}^{\mu(p)} - \lim_{s\to\infty} \|g_\sigma - g_{\sigma,n_s}\|_{L_p[-M,M]}^{\mu(p)}$$
(5.2.4) $\qquad = \|f - g_\sigma\|_{L_p[-M,M]}^{\mu(p)}.$

Letting now $M \to \infty$ in (5.2.4), we obtain (5.2.1) from (5.2.4). ∎

Next we show that a best approximant to a function of polynomial growth in $L_p(\mathbb{R})$ has a polynomial growth as well.

PROPOSITION 5.2.2. *If $f \in M_{A,N}$ and $A_\sigma(f, L_p) < \infty$, $0 < p < \infty$, then for g_σ, satisfying (5.2.1), we have $g_\sigma \in M_{C(C_1+A),N}$, where $C = C(\sigma)$ and $C_1 = A_\sigma(f, L_p)$.*

PROOF. Note first that

$$|E_1| := \bigl|\{x \in \mathbb{R} : |f(x) - g_\sigma(x)| > \|f - g_\sigma\|_{L_p(\mathbb{R})}\}\bigr| \le 1.$$

Then for every $x \in E := \mathbb{R} \setminus E_1$, we have

$$|g_\sigma(x)| \le \bigl(A_\sigma(f, L_p) + A\bigr)(1 + |x|)^N.$$

Using Proposition 5.1.1(b), we conclude that $g_\sigma \in M_{C(C_1+A),N}$. ∎

Finally, we discuss conditions that guarantee continuity of $A_\sigma(f, L_p)$ in $\sigma > 0$.

PROPOSITION 5.2.3. *If for some $\sigma_0 > 0$,*

(5.2.5) $\qquad A_{\sigma_0}(f, L_p) < \infty,$

then the following statements hold:

(a) For every $\sigma \ge \sigma_0$ and $0 < p \le \infty$,

(5.2.6) $\qquad \lim_{\tau \to \sigma+} A_\tau(f, L_p) = A_\sigma(f, L_p).$

(b) For every $\sigma > \sigma_0$ and $0 < p < \infty$,

(5.2.7) $\qquad \lim_{\tau \to \sigma-} A_\tau(f, L_p) = A_\sigma(f, L_p).$

PROOF. (a) Condition (5.2.5) implies that there exists $g_0 \in B_{\sigma_0}$ such that $f - g_0 \in L_p(\mathbb{R})$ and for $\sigma \geq \sigma_0$,
$$A_\sigma(f, L_p) = A_\sigma(f - g_0, L_p).$$
Hence without loss of generality we can assume that $f \in L_p(\mathbb{R})$. Next by Proposition 5.2.1, if $\{\varepsilon_n\}_{n=1}^\infty$ is a sequence of positive numbers with $\lim_{n \to \infty} \varepsilon_n = 0$, then there exists a sequence of functions $g_{\sigma,n} \in B_{\sigma+\varepsilon_n}$, satisfying

(5.2.8) $\qquad \|f - g_{\sigma,n}\|_{L_p(\mathbb{R})} = A_{\sigma+\varepsilon_n}(f, L_p), \qquad n = 1, 2, \ldots.$

Then using (1.3.1), (1.3.4), and (5.2.8), we get
$$\|g_{\sigma,n}\|_{L_p(\mathbb{R})} \leq 2^{1/\mu(p)} \|f\|_{L_p(\mathbb{R})}, \qquad n = 1, 2, \ldots.$$
Applying Proposition 5.1.2(b), we obtain that there exist a subsequence $\{g_{\sigma,n_s}\}_{s=1}^\infty$ and $g_\sigma \in \bigcap_{n=1}^\infty B_{\sigma+\varepsilon_n} = B_\sigma$ such that

(5.2.9) $\qquad \lim_{s \to \infty} g_{\sigma,n_s} = g_\sigma$

uniformly in any interval $[-M, M]$. Further, it follows from (5.2.8) and (5.2.9)

$$\begin{aligned}
A_\sigma^{\mu(p)}(f, L_p) &\leq \lim_{M \to \infty} \|f - g_\sigma\|_{L_p[-M,M]}^{\mu(p)} \\
&\leq \lim_{M \to \infty} \left(\liminf_{s \to \infty} \|f - g_{\sigma,n_s}\|_{L_p[-M,M]}^{\mu(p)} \right. \\
&\qquad \left. + \lim_{s \to \infty} \|g_\sigma - g_{\sigma,n_s}\|_{L_p[-M,M]}^{\mu(p)} \right) \\
&\leq \liminf_{n \to \infty} A_{\sigma+\varepsilon_n}^{\mu(p)}(f, L_p) \leq \limsup_{n \to \infty} A_{\sigma+\varepsilon_n}^{\mu(p)}(f, L_p) \leq A_\sigma^{\mu(p)}(f, L_p).
\end{aligned}$$

This yields (5.2.6).

(b) Without loss of generality, we assume again that $f \in L_p(\mathbb{R})$, $0 < p < \infty$. Then by Proposition 5.2.1, there exists $g_\sigma \in B_\sigma$ such that

(5.2.10) $\qquad \|f - g_\sigma\|_{L_p(\mathbb{R})} = A_\sigma(f, L_p).$

Next given $\delta \in (0, \sigma/2)$, we consider the function
$$g_{\sigma,\delta}(x) := g_\sigma\left(\frac{\sigma - \delta}{\sigma} x\right).$$
Then $g_{\sigma,\delta} \in B_{\sigma-\delta}$ and by (5.2.10),

$$\begin{aligned}
A_{\sigma-\delta}^{\mu(p)}(f, L_p) &\leq \|f - g_{\sigma,\delta}\|_{L_p(\mathbb{R})}^{\mu(p)} \leq \|f - g_\sigma\|_{L_p(\mathbb{R})}^{\mu(p)} + \|g_\sigma - g_{\sigma,\delta}\|_{L_p(\mathbb{R})}^{\mu(p)} \\
&\leq A_\sigma^{\mu(p)}(f, L_p) + \|g_\sigma - g_{\sigma,\delta}\|_{L_p[-M,M]}^{\mu(p)}
\end{aligned}$$

(5.2.11) $\qquad\qquad\qquad + \|g_\sigma - g_{\sigma,\delta}\|_{L_p(|x| \geq M)}^{\mu(p)},$

where $M > 0$.

We can choose M large enough such that

$$\|g_\sigma - g_{\sigma,\delta}\|_{L_p(|x| \geq M)}^{\mu(p)} \leq \|g_\sigma\|_{L_p(|x| \geq M)}^{\mu(p)} + \|g_{\sigma,\delta}\|_{L_p(|x| \geq M)}^{\mu(p)}$$

(5.2.12) $\qquad\qquad\qquad \leq \left(1 + 2^{\mu(p)/p}\right) \|g_\sigma\|_{L_p(|x| \geq M/2)}^{\mu(p)} < \varepsilon.$

5.2. APPROXIMATION PROPERTIES OF ENTIRE FUNCTIONS OF EXPONENTIAL TYPE

Further, we choose δ small enough such that

$$(5.2.13) \quad \|g_\sigma - g_{\sigma,\delta}\|_{L_p[-M,M]}^{\mu(p)} \le (2M)^{\mu(p)/p} \max_{x \in [-M,M]} \left| g_\sigma(x) - g_\sigma\left(\frac{\sigma-\delta}{\sigma}x\right) \right|^{\mu(p)} < \varepsilon.$$

Inequalities (5.2.11)–(5.2.13) imply

$$A_{\sigma-\delta}^{\mu(p)}(f, L_p) \le A_\sigma^{\mu(p)}(f, L_p) + 2\varepsilon.$$

Thus (5.2.7) holds. ∎

Remarks

(a) Propositions 5.2.1 and 5.2.3 for $1 \le p \le \infty$ are proved in [114] and [4], respectively. More general versions of Propositions 5.2.1–5.2.3 were discussed in [47].

(b) Proposition 5.2.3 shows that $A_\sigma(f, L_p)$ is continuous in $\sigma > \sigma_0$ if $p \in (0, \infty)$. The situation is different if $p = \infty$. For example, for $f(x) = \sin x$, $\sigma = 1$, we have $A_{\sigma+}(f, L_\infty) = A_\sigma(f, L_\infty) < A_{\sigma-}(f, L_\infty)$. Moreover, Gordon [58] showed that for every nonincreasing function $A(\sigma): [0, \infty) \to [0, \infty)$, satisfying the conditions $A(\sigma) = A(\sigma+)$, $\sigma \in [0, \infty)$, and $\lim_{\sigma \to \infty} A(\sigma) = 0$, there exists $f_A \in L_\infty(\mathbb{R})$ such that $A_\sigma(f_A, L_\infty) = A(\sigma)$ for all $\sigma \in [0, \infty)$.

(c) Each of the following conditions on a continuous f with $A_{\sigma_0}(f, L_\infty) < \infty$ is sufficient for $A_\sigma(f, L_\infty)$ to be continuous in $\sigma > \sigma_0$:

(i) $\limsup_{t \to 1} \sup_{x \in \mathbb{R}} |f(tx) - f(x)| = 0$ (see [4]),

(ii) f is λ-homogeneous on $\mathbb{R}$, $\lambda > 0$ (cf. [40]).

CHAPTER 6

Polynomial Interpolation and Approximation of Entire Functions of Exponential Type

In this chapter, we obtain estimates for the error of weighted approximation of an entire function g of exponential type $\sigma < b_n$ by the Lagrange interpolation polynomials $P_{n-1} \in \mathcal{P}_{n-1}$ to g at the zeros $\{x_{k,n}^{(q)}\}_{k=1}^{n}$ of the extremal polynomials $T_{n,q}$, $1 < q < \infty$.

In order to discuss both cases of a bounded and unbounded interval $I = (-c,c)$ simultaneously, throughout this chapter we assume that for $0 < c \leq \infty$, a weight $W(x) = \exp(-Q(x)) \in \mathcal{F}_C$ is extended from $(-c,c)$ to $\mathbb{R}$ by setting $W(x) = 0$ for $|x| \geq c$. The extended weight (we denote it again by W) is continuous on $\mathbb{R}$, by properties (a) and (d) of Definition 1.4.2.

Note that by statements (c) and (d) of Proposition 4.2.1, all interpolation nodes $\{x_{k,n}^{(q)}\}_{k=1}^{n}$ are contained in the interval

(6.0.14) $$I_n := [-a_n(1+\delta_n), a_n(1+\delta_n)],$$

where $\{\delta_n\}_{n=1}^{\infty}$ is a sequence of numbers, satisfying either

(6.0.15) $$C/(n+1) \leq \delta_n < 1, \quad n > C_1, \quad W \in \mathcal{F}_C,$$

or

(6.0.16) $$C(\log n/n)^{2/3} \leq \delta_n < 1, \quad n > C_2, \quad W \in \mathcal{F}(C^3).$$

To prove Theorem 2.2.1 and some other results, we shall use the inequalities

(6.0.17) $E_n(g, L_{p,W}(I)) \leq \|g - P_{n-1}\|_{L_{p,W}(I)} \leq 2^{1/\mu(p)-1}$
$$\times \left(\|g - P_{n-1}\|_{L_{p,W}(I_n)} + \|g - P_{n-1}\|_{L_{p,W}(I \setminus I_n)}\right) := 2^{1/\mu(p)-1}(J_1 + J_2),$$

where $g \in B_\sigma$, $0 < p \leq \infty$, I_n is defined in (6.0.14), and δ_n satisfies either (6.0.15) or (6.0.16), $n = 1, 2, \ldots$.

Note that if $a_n(1+\delta_n) > c$, then by the assumption,

$$J_1 = \|g - P_{n-1}\|_{L_{p,W}(I)}, \quad J_2 = 0.$$

To prove Theorem 2.2.2, we shall need to estimate J_1 only, while for the proof of Theorem 2.2.1 and some other results estimates of both J_1 and J_2 are needed.

6.1. Interpolation on the Interval $I_n = [-a_n(1+\delta_n), a_n(1+\delta_n)]$

We first obtain an estimate for J_1 when g is a function from B_σ.

PROPOSITION 6.1.1. *Let $W \in \mathcal{F}(C^2)$ and let $\{\delta_n\}_{n=1}^{\infty}$ satisfy (6.0.15). Then there exists $\mu_1 \in (0, 1/3)$ such that for any $g \in B_\sigma$ and for every R satisfying*

(6.1.1) $$R \geq \max(1 + 5\sqrt{\delta_n}, 1 + C_1 n^{-\mu_1}), \quad n > C_2,$$

the following inequality holds

$$(6.1.2) \qquad J_1 \leq C n^{\gamma_1} R^{-n+1} e^{-H(i/R)} \max_{|w|=R} \left| g\left(\frac{a_n}{2}\left(w + \frac{1}{w}\right)\right) \right|,$$

where H is defined in (3.0.7) and the constants C, C_1, C_2, and $\gamma_1 \geq 0$ are independent of n, R, σ, and g.

PROOF. Let us consider the ellipse $E(a_n, R)$ with foci at $\pm a_n$ and with the sum of its semi-axes equal to $a_n R$. It follows from (6.0.15) and (6.1.1) that the horizontal semi-axis s of $E(a_n, R)$ satisfies the relations

$$s = \frac{a_n}{2}\left(R + \frac{1}{R}\right) = a_n\left(1 + \frac{(R-1)^2}{2R}\right) \geq a_n\left(1 + \frac{25\delta_n}{2(1+5\delta_n)}\right) \geq a_n(1 + 2\delta_n).$$

This immediately implies the following properties of $E(a_n, R)$:

(i) $I_n \subseteq E(a_n, R)$;
(ii) $E(a_n, R)$ contains all nodes $\{x_{k,n}^{(q)}\}_{k=1}^n$;
(iii)

$$(6.1.3) \qquad \min_{x \in I_n} \min_{z \in E(a_n, R)} |x - z| \geq \delta_n a_n.$$

Properties (i) and (ii) show that the Hermite error formula for Lagrange interpolation [121] can be applied to $g - P_{n-1}$:

$$g(x) - P_{n-1}(x) = \frac{T_{n,q}(x)}{2\pi i} \int_{E(a_n, R)} \frac{g(z)}{(z-x)T_{n,q}(z)} dz.$$

Hence using (6.1.3), we obtain

$$(6.1.4) \qquad \|g - P_{n-1}\|_{L_{p,W}(I_n)} \leq \frac{\|T_{n,q}\|_{L_{p,W}(I_n)}}{2\pi \delta_n a_n} \int_{E(a_n, R)} \frac{|g(z)|}{|T_{n,q}(z)|} |dz|.$$

Next, we use Nikolskii-type inequality (4.1.7) for the extremal polynomials

$$(6.1.5) \qquad \|T_{n,q}\|_{L_{p,W}(I_n)} \leq \|T_{n,q}\|_{L_{p,W}(I)} \leq C n^{2|1/p - 1/q|} \|T_{n,q}\|_{L_{q,W}(I)},$$

where $p \in (0, \infty]$ and $q \in (1, \infty)$. Then we deduce the following estimate from (6.1.4) and (6.1.5):

$$(6.1.6) \qquad \|g - P_{n-1}\|_{L_{p,W}(I_n)} \leq \frac{Cn^{2|1/p - 1/q|}}{\delta_n a_n} \int_{E(a_n, R)} \frac{|g(z)|}{|P_{n,q}(z)|} |dz|$$

where $P_{n,q}$ is the normalized extremal polynomial defined in (4.2.3).

Further making the substitution $z = (a_n/2)(w + w^{-1})$ in (6.1.6) and taking account of the inequality

$$(6.1.7) \qquad \max_{|w|=R} |1 - w^{-2}| \leq 2,$$

we obtain from (6.1.6) and (6.0.15)

$$(6.1.8) \quad \|g - P_{n-1}\|_{L_{p,W}(I_n)}$$
$$\leq Cn^{2|1/p-1/q|+1} \max_{|w|=R} \left|g\left(\frac{a_n}{2}\left(w + \frac{1}{w}\right)\right)\right| \int_{|w|=R} \frac{|dw|}{|P_{n,q}\left(\frac{a_n}{2}\left(w + \frac{1}{w}\right)\right)|}.$$

Furthermore, we recall that by Proposition 4.2.3, there exists $\mu_1 \in (0, 1/3)$ such that for any $R = |w| \geq 1 + C_2 n^{-\mu_1}$, $q \in (1, \infty)$,

$$\left| P_{n,q}\left(\frac{a_n}{2}\left(w + \frac{1}{w}\right)\right) \right| \geq C a_n^{-1/q} R^n |\exp(H(1/w))|. \tag{6.1.9}$$

Since $a_n \leq Cn$ by (1.4.19), we have from (6.1.8) and (6.1.9)

$$\|g - P_{n-1}\|_{L_{p,W}(I_n)} \leq C n^{2|1/p - 1/q| + 1 + 1/q} R^{-n+1}$$
$$\times \max_{|w|=R} \exp(-\operatorname{Re} H(1/w)) \max_{|w|=R} \left| g\left(\frac{a_n}{2}\left(w + \frac{1}{w}\right)\right) \right|. \tag{6.1.10}$$

Thus (6.1.2) follows from (6.1.10) and Proposition 3.1.3. ∎

Next, we prove two versions of Proposition 6.1.1 for $W \in \mathcal{F}(C^3)$ and $W = W_1$.

PROPOSITION 6.1.2. *Let $W \in \mathcal{F}(C^3)$ and let $\{\delta_n\}_{n=1}^\infty$ satisfy (6.0.16). Then for any $\varepsilon > 0$, each $g \in B_\sigma$, and every R satisfying*

$$R \geq \max(1 + 5\sqrt{\delta_n}, 1 + \varepsilon), \qquad n > C_2, \tag{6.1.11}$$

estimate (6.1.2) holds.

PROOF. Following the proof of Proposition 6.1.1, we see that inequality (6.1.4) holds for $R \geq 1 + 5\sqrt{\delta_n}$. Then by Proposition 4.1.1(b), we get

$$\|g - P_{n-1}\|_{L_{p,W}(I_n)} \leq \frac{C n^{1/q + 2/\min(1,p)}}{\delta_n a_n} \int_{E(a_n, R)} \frac{|g(z)|}{|P_{n,q}(z)|} |dz|. \tag{6.1.12}$$

Next making the substitution $z = (a_n/2)(w + w^{-1})$ in (6.1.12) and using (6.0.16), (6.1.7), we obtain

$$\|g - P_{n-1}\|_{L_{p,W}(I_n)} \leq C n^{1/q + 2/\min(1,p) + 2/3} \max_{|w|=R} \left| g\left(\frac{a_n}{2}\left(w + \frac{1}{w}\right)\right) \right|$$
$$\times \int_{|w|=R} \frac{|dw|}{|P_{n,q}(\frac{a_n}{2}(w + \frac{1}{w}))|}. \tag{6.1.13}$$

Since by Proposition 4.2.2, (6.1.9) holds for $R = |w| \geq 1 + \varepsilon$ if $W \in \mathcal{F}(C^3)$, we have from (6.1.9), (6.1.13) and Proposition 3.1.3

$$\|g - P_{n-1}\|_{L_{p,W}(I_n)} \leq C a_n^{-1/q} n^{1/q + 2/\min(1,p) + 2/3}$$
$$\times R^{-n+1} e^{-H(i/R)} \max_{|w|=R} \left| g\left(\frac{a_n}{2}\left(w + \frac{1}{w}\right)\right) \right|.$$

Thus to complete the proof of the proposition, it remains to show that for $W \in \mathcal{F}(C^3)$,

$$a_n \leq C n^{1/C_1}, \tag{6.1.14}$$

where C_1 is the constant from Definition 1.4.6. Indeed, we first note that (1.4.8) and (1.4.12) imply $\lim_{n \to \infty} a_n = \infty$. Then using the estimate [84, Lemma 6.2(a)]

$$Q'(x) \geq Q'(1) x^{C_1 - 1}, \qquad x \in (1, \infty),$$

we get from (1.4.12) for n large enough
$$n \geq \frac{2}{\pi a_n} \int_1^{a_n} xQ'(x)\,dx \geq Ca_n^{C_1}.$$

Hence (6.1.14) follows. ∎

The weight $W = \exp(-|x|)$ belongs to $\mathcal{F}(C^3)$ but does not belong to $\mathcal{F}(C^2)$. It turns out that for this weight inequality (6.1.2) holds under weaker conditions than in Proposition 6.1.2.

PROPOSITION 6.1.3. *Let $W = W_1$. Then for any $\varepsilon_1 \in (0, 1/3)$, each $g \in B_\sigma$, and every R, satisfying*
$$R \geq \max\bigl(1 + 5\sqrt{\delta_n},\ 1 + C_1 n^{-1/3+\varepsilon_1}\bigr), \qquad n > C_2,$$
inequality (6.1.2) holds. Here $\{\delta_n\}_{n=1}^\infty$ satisfies (6.0.16), and $P_{n-1} \in \mathcal{P}_{n-1}$ in (6.1.2) is the Lagrange interpolation polynomial to g at the zeros $\{x_{k,n}^{(2)}\}_{k=1}^n$ of the orthogonal polynomial $T_{n,2}$ associated with W_1^2.

PROOF. Since $W_1 \in \mathcal{F}(C^3)$, (6.1.13) holds for $q = 2$. Next, we make use of estimate (4.2.24) for the orthonormal polynomial $P_{n,2}$ associated with W_1^2 that holds for $R \geq 1 + C_1 n^{-1/3+\varepsilon_1}$, by Proposition 4.2.5. Thus relations (6.1.13), (4.2.24), (4.2.25) and Proposition 3.1.3 yield (6.1.2). ∎

The following proposition is an immediate consequence of Propositions 6.1.1, 5.1.3 and inequality (1.4.19):

PROPOSITION 6.1.4. *Let $W \in \mathcal{F}(C^2)$. Then there exists $\mu_1 \in (0, 1/3)$ such that for $n > C_1$ and every $r \in (0, 1)$, satisfying*

$$\text{(6.1.15)} \qquad r \leq \min\left(1 - \frac{5\sqrt{\delta_n}}{1 + 5\sqrt{\delta_n}},\ 1 - C_3 n^{-\nu_1}\right),$$

the following statements hold:

(a) *If a function $g \in B_\sigma$ satisfies (2.2.2), then*

$$\text{(6.1.16)} \qquad J_1 \leq ACn^{\gamma_2} r^{-N-1} \exp\bigl(S_\sigma(r)\bigr),$$

where

$$\text{(6.1.17)} \qquad S_\sigma(r) := \frac{a_n \sigma}{2}\left(\frac{1}{r} - r\right) - H(ir) + n\log r.$$

(b) *If $g \in B_\sigma$, then for every $\varepsilon > 0$ there exists $C(g, \varepsilon)$ such that*

$$\text{(6.1.18)} \qquad J_1 \leq CC(g, \varepsilon) n^{\gamma_2} r^{-1} \exp\bigl(G_{\sigma(1+\varepsilon)}(r)\bigr),$$

where

$$\text{(6.1.19)} \qquad G_\sigma(r) := \frac{a_n \sigma}{2}\left(\frac{1}{r} + r\right) - H(ir) + n\log r.$$

6.2. Interpolation on $I \setminus I_n$

Here, we estimate the "tail" J_2 in (6.0.17) for $W \in \mathcal{F}_C$ and $W \in \mathcal{F}(C^3)$.

PROPOSITION 6.2.1. *Let $W \in \mathcal{F}_C$. Then there exists a number $\varepsilon_1 > 0$ such that for any δ_n, satisfying (6.0.15), and any $g \in B_\sigma$, $\sigma > 0$, satisfying (2.2.2), we have*

$$(6.2.1) \qquad J_2 \leq C n^{m_1} \big(\exp(-C_2 n \delta_n^{3/2}) + \exp(-C_3 n^{\varepsilon_1})\big)(J_1 + A),$$

where C, C_2, C_3, and $m_1 \geq 0$ are independent of n, σ, and g.

PROOF. We first prove that for any $g \in B_\sigma$,

$$(6.2.2) \qquad \begin{aligned} J_2 &\leq C_4 \big(\|g\|_{L_{p,W}(I \setminus I_n)} + n^{m_2}\big(\exp(-C_2 n \delta_n^{3/2}) + \exp(-C_3 n^{\varepsilon_2})\big)\big) \\ &\quad \times \big(J_1 + \|g\|_{L_{p,W}(I)}\big). \end{aligned}$$

Indeed, we have

$$(6.2.3) \qquad J_2 \leq 2^{1/\mu(p)-1}\big(\|g\|_{L_{p,W}(I \setminus I_n)} + \|P_{n-1}\|_{L_p(I \setminus I_n)}\big).$$

Let first $p \in (0, \infty)$. Then using Propositions 4.3.1, 4.3.2(a), 4.3.2(e) and inequality (1.4.19), we get

$$\|P_{n-1}\|_{L_{p,W}(I \setminus I_n)}$$
$$\leq \left(\int_{|x| \geq a_n(1+\delta_n)} \exp(pU_n(x))\,dx\right)^{1/p} \|P_{n-1}\|_{L_{\infty,W}[-a_n, a_n]}$$
$$\leq \left(\int_{a_n(1+\delta_n) \leq |x| \leq a_{2n}} \exp(pU_n(x))\,dx \right.$$
$$\left. + \sum_{r=2n}^{\infty} \int_{a_r}^{a_{r+1}} \exp(pU_n(x))\,dx\right)^{1/p} \|P_{n-1}\|_{L_{\infty,W}[-a_n, a_n]}$$
$$\leq \left(\int_{a_n(1+\delta_n) \leq |x| \leq a_{2n}} \exp\left(-C_5 p n \left(\frac{|x|}{a_n} - 1\right)^{3/2}\right) dx \right.$$
$$\left. + \sum_{r=2n}^{\infty} \exp(-C_6 p r^{\varepsilon_1})(a_{r+1} - a_r)\right)^{1/p} \|P_{n-1}\|_{L_{\infty,W}[-a_n, a_n]}$$
$$\leq C\big(n \exp(-C_2 p n \delta_n^{3/2}) + n^{2-\varepsilon_1} \exp(-C_7 p (2n)^{\varepsilon_2})\big)^{1/p} \|P_{n-1}\|_{L_{\infty,W}[-a_n, a_n]}.$$

Hence by Nikolskii-type inequality (4.1.3) and by estimate (1.4.28), we have

$$(6.2.4) \qquad \begin{aligned} \|P_{n-1}\|_{L_{p,W}(I \setminus I_n)} &\leq C n^{m_2}\big(\exp(-C_2 n \delta_n^{3/2}) \\ &\quad + \exp(-C_3 n^{\varepsilon_2})\big) \|P_{n-1}\|_{L_{p,W}[-a_n, a_n]}. \end{aligned}$$

Then taking account of the inequality

$$\|P_{n-1}\|_{L_{p,W}[-a_n, a_n]} \leq 2^{1/\mu(p)-1}\big(\|g - P_{n-1}\|_{L_{p,W}(I_n)} + \|g\|_{L_{p,W}(I)}\big),$$

we obtain (6.2.2) for $p \in (0, \infty)$ from (6.2.3) and (6.2.4).

If $p = \infty$, then by Proposition 4.3.1, 4.3.2(a), and 4.3.2(c),

$$(6.2.5) \qquad \begin{aligned} \|P_{n-1}\|_{L_{\infty,W}(I \setminus I_n)} &\leq \exp(U_n(a_n(1+\delta_n)))\|P_{n-1}\|_{L_{\infty,W}[-a_n, a_n]} \\ &\leq \exp(-C_2 n \delta_n^{3/2})\|P_{n-1}\|_{L_{\infty,W}[[-a_n, a_n]}. \end{aligned}$$

Thus (6.2.2) for $p = \infty$ follows from (6.2.3) and (6.2.5).

Next if $g \in B_\sigma$ and g satisfies (2.2.2), then for $0 < p \leq \infty$

$$\|g\|_{L_{p,W}(I \setminus I_n)} \leq An^m \left(\int_{|x| \geq a_n(1+\delta_n)} (1 + |x|)^{Np} e^{-pQ(x)} dx \right)^{1/p}$$

(6.2.6)
$$\leq ACn^m \exp\left(-\frac{1}{2}Q(a_n)\right) \leq ACn^m (-C_3 n^{\varepsilon_3}),$$

by (1.4.21).

Since

(6.2.7) $$\|g\|_{L_{p,W}(I)} \leq ACn^m,$$

we obtain (6.2.1) from (6.2.2), (6.2.6), and (6.2.7) with $\varepsilon_1 = \min(\varepsilon_2, \varepsilon_3)$ and $m_1 = \max(m, m_2)$. ∎

PROPOSITION 6.2.2. *Let $W \in \mathcal{F}(C^3)$ and $p = \infty$. Then for any δ_n, satisfying (6.0.16), and any $g \in B_\sigma$,*

(6.2.8) $$J_2 \leq Cn^{m_1} \exp(-C_2 n \delta_n^{3/2})(J_1 + \|g\|_{L_\infty, W(\mathbb{R})}) + \|g\|_{L_\infty, W(|x| \geq a_n)}.$$

PROOF. The proof is similar to that of Proposition 6.2.1. We first need an analogue of (6.2.5) for $W \in \mathcal{F}(C^3)$. It was shown in [84, Eq. (5.54)] that for $W \in \mathcal{F}(C^3)$ and $\varepsilon \in (0, 1)$,

(6.2.9) $$|n^{-1} U_n(a_n(1 + \varepsilon)) + C_0 \varepsilon^{3/2}| \leq C\varepsilon^{5/2},$$

where C_0 is an absolute constant and C is independent of n and ε.

Next, estimate (4.3.2) holds for $W \in \mathcal{F}(C^3)$ as well (see [84, Eq. (7.5)]), so we obtain from (4.3.2) and (6.2.9) for $\varepsilon = \delta_n$

(6.2.10) $$\|P_{n-1}\|_{L_\infty, W(\mathbb{R} \setminus I_n)} \leq \exp(-C_2 n \delta_n^{3/2}) \|P_{n-1}\|_{L_\infty, W[-a_n, a_n]}.$$

Then using (6.2.10), we have

$$\begin{aligned} J_2 &\leq \|P_{n-1}\|_{L_\infty, W(\mathbb{R} \setminus I_n)} + \|g\|_{L_\infty, W(\mathbb{R} \setminus I_n)} \\ &\leq \exp(-C_2 n \delta_n^{3/2}) \|P_{n-1}\|_{L_\infty, W[-a_n, a_n]} + \|g\|_{L_\infty, W(\mathbb{R} \setminus I_n)} \\ &\leq \exp(-C_2 n \delta_n^{3/2})(J_1 + \|g\|_{L_\infty, W(\mathbb{R})}) + \|g\|_{L_\infty, W(\mathbb{R} \setminus I_n)}. \end{aligned}$$

This yields (6.2.8). ∎

6.3. Proof of Theorem 2.2.1

We first study properties of the function $S_\sigma(r)$ defined in (6.1.17). It is obvious that S_σ is an infinitely differentiable function in $r \in (0, 1)$. Also, S_σ is real-valued for $r > 0$, by Proposition 3.1.2(a). Next, it is easy to verify the identity

(6.3.1) $$\frac{dS_\sigma(r)}{dr} = \frac{a_n(1 + r^2)}{2r^2}(h(r) - \sigma),$$

where h is defined in (3.2.1).

Since by statements (c) and (d) of Proposition 3.2.1, h is a positive, increasing and continuous function on $(0, 1)$ and $h(0) = 0$, $h(1) = b_n$, we have from (6.3.1) that for $0 \leq \sigma \leq b_n$

(6.3.2) $$\min_{r \in (0,1)} S_\sigma(r) = S_\sigma(r(\sigma)) = \phi(r(\sigma)),$$

where ϕ is defined in (3.3.1) and $r(\sigma) = h^{(-1)}(\sigma)$ is the unique solution to the equation $h(r) = \sigma$. Since $r(\sigma)$ is increasing in $\sigma \in [0, b_n]$ and by Proposition 3.3.1, $\phi(r)$ is increasing in $r \in (0, 1)$, the following estimate holds:

$$\min_{r \in (0,1)} S_\sigma(r) \leq \phi\big(r(b_n(1 - \lambda_n))\big), \qquad \sigma \in \big[0, b_n(1 - \lambda_n)\big], \tag{6.3.3}$$

where λ_n is a number from $(0, 1)$.

Further, using Proposition 3.3.2, we obtain from (6.3.2) and (6.3.3)

$$\max_{\sigma \in [0, b_n(1-\lambda_n)]} S_\sigma(r(\sigma)) \leq -(n/36)\big(1 - r(b_n(1 - \lambda_n))\big)^3. \tag{6.3.4}$$

Let us choose

$$\lambda_n = 1 - h\big(1 - C_3 n^{-\mu_1}\big)/b_n, \qquad n > n_0, \tag{6.3.5}$$

which is equivalent to

$$r\big(b_n(1 - \lambda_n)\big) = 1 - C_3 n^{-\mu_1}, \qquad n > n_0, \tag{6.3.6}$$

where $\mu_1 \in (0, 1/3)$ and $C_3 > 0$ are the same constants as in Proposition 6.1.4 and $n_0 > C_1$. Then (6.3.4) and (6.3.6) imply the estimate

$$\max_{\sigma \in [0, b_n(1-\lambda_n)]} S_\sigma(r(\sigma)) = S_\sigma\big(r(b_n(1-\lambda_n))\big) \leq -C n^{1 - 3\mu_1}, \tag{6.3.7}$$

where C is an absolute constant and λ_n is given by (6.3.5).

Let now g be an entire function of exponential type $\sigma < b_n$, satisfying (2.2.2). We first give an exponential estimate for J_1 in (6.0.17).

Let us choose δ_n in (6.0.14) by

$$\delta_n = C n^{-2\mu_1} \tag{6.3.8}$$

and choose r in (6.1.16) by

$$r = r\big(b_n(1 - \lambda_n)\big) = 1 - C_3 n^{-\mu_1}. \tag{6.3.9}$$

Then δ_n satisfies (6.0.15) for n large enough, and r satisfies (6.1.15) if C in (6.3.8) is small enough. Using Proposition 6.1.4(a) and inequality (6.3.7), we have for any $\sigma \in [0, b_n(1 - \lambda_n)]$

$$\begin{aligned} J_1 &= \|g - P_{n-1}\|_{L_{p,W}[-a_n(1+Cn^{-2\mu_1}), a_n(1+Cn^{-2\mu_1})]} \\ &\leq ACn^{\gamma_2}\big(1 - C_3 n^{-\mu_1}\big)^{-N-1} \exp\Big(\max_{\sigma \in (0, b_n[1-\lambda_n)]} S_\sigma(r(\sigma))\Big) \\ &\leq ACn^{\gamma_2} \exp\big(-C_1 n^{1-3\mu_1}\big). \end{aligned} \tag{6.3.10}$$

Finally, it follows from Proposition 3.2.3 and (6.3.5) that

$$\lambda_n \leq C_4 n^{-\frac{2(\Lambda-1)}{\Lambda+1}\mu_1}. \tag{6.3.11}$$

Thus (6.3.10) and (6.3.11) yield the estimate

$$J_1 \leq ACn^{\gamma_2} \exp\big(-C_1 n^{1-3\mu_1}\big), \tag{6.3.12}$$

if $\sigma \in \Big(0, b_n\big(1 - C_4 n^{-\frac{2(\Lambda-1)}{\Lambda+1}\mu_2}\big)\Big]$.

To estimate J_2, we use Proposition 6.2.1 and relations (6.3.8), (6.3.12)

$$\begin{aligned} J_2 &\leq \|g - P_{n-1}\|_{L_{p,W}(|x| \geq a_n(1+Cn^{-2\mu_1}))} \\ &\leq ACn^{m_1}\Big(\exp\big(-C_5 n^{1-3\mu_1}\big) + \exp\big(-C_6 n^{-\varepsilon_1}\big)\Big). \end{aligned} \tag{6.3.13}$$

Therefore, we have from (6.0.17), (6.3.12), and (6.3.13)

$$E_n(g, L_{p,W}(I)) \leq \|g - P_{n-1}\|_{L_{p,W}(I)} \leq C_2 A n^\gamma \exp(-C_3 n^{\delta_2}), \tag{6.3.14}$$

where $\sigma = (0, b_n(1 - C_1 n^{-\delta_1})]$, $\delta_1 = \frac{2(\Lambda-1)}{\Lambda+1}\mu_1$, $\delta_2 = \min(1 - 3\mu_1, \varepsilon_1)$, $\mu_1 \in (0, 1/3)$, and $\gamma = \max(\gamma_2, m_1)$.

This completes the proof of Theorem 2.2.1. ∎

6.4. Proof of Theorem 2.2.2

To prove the theorem, we need certain properties of the function $G_\sigma(r)$ defined in (6.1.19).

We first note that $G_\sigma(r)$ is a real-valued and infinitely differentiable function in $r \in (0, 1)$. Next, we have from (6.1.19) and (3.2.1)

$$\frac{dG_\sigma(r)}{dr} = \frac{a_n(1-r^2)}{2r^2}\left(h(r)\frac{1+r^2}{1-r^2} - \sigma\right). \tag{6.4.1}$$

Since $\varphi(r) := h(r)(1+r^2)/(1-r^2)$ is increasing and positive in $(0, 1)$, and $\varphi(0) = 0$, $\varphi(1) = \infty$, we obtain from (6.4.1) that for $0 < \sigma < \infty$,

$$\min_{r \in (0,1)} G_\sigma(r) = G_\sigma(r^*(\sigma)) = \phi_1((r^*(\sigma)), \tag{6.4.2}$$

where ϕ_1 is defined in (3.3.16), and $r^*(\sigma) := \varphi^{(-1)}(\sigma)$ is the unique solution to the equation

$$h(r)\frac{1+r^2}{1-r^2} = \sigma. \tag{6.4.3}$$

By statements (b) and (d) or Proposition 3.3.6, we have that ϕ_1 is increasing in $r \in (0, 1)$ and there exists the unique solution $r_0 \in (0, 1)$ to the equation

$$\phi_1(r) = 0.$$

Further, Proposition 3.3.8 shows that

$$0.24 < r_0 < 0.60. \tag{6.4.4}$$

Then $\min_{r \in (0,1)} G_\sigma(r) < 0$ for $\sigma \in (0, \sigma_0)$, where

$$\sigma_0 := \frac{h(r_0)(1 + r_0^2)}{1 - r_0^2}.$$

Moreover, by Proposition 3.3.7 and (6.4.2),

$$\min_{r \in (0,1)} G_\sigma(r) = G_\sigma(r^*(\sigma)) \leq -n(r_0 - r^*(\sigma)), \qquad 0 < \sigma \leq \sigma_0. \tag{6.4.5}$$

Next, for a fixed number $\delta \in (0, r_0)$, we set

$$\varepsilon_1 = 1 - \frac{h(r_0 - \delta)(1 + (r_0 - \delta)^2)(1 - r_0^2)}{h(r_0)(1 + r_0^2)(1 - (r_0 - \delta)^2)}. \tag{6.4.6}$$

Then inequalities (6.4.4) show that we can estimate ε_1 by the right inequality in (3.2.23) of Proposition 3.2.5 with $C_0 = 0.04$, $C_1 = 0.60$, $r_1 = r_0 - \delta$, $r_2 = r_0$, and $\delta \in (0, 0.2]$. Namely,

$$0 < \varepsilon_1 < C\delta, \tag{6.4.7}$$

where C is an absolute constant. Since (6.4.6) is equivalent to

$$r_0 - r^*(\sigma_0(1 - \varepsilon_1)) = \delta, \tag{6.4.8}$$

we obtain from (6.4.5), (6.4.7), and (6.4.8)

(6.4.9) $$\max_{0\leq\sigma\leq\sigma_0(1-\varepsilon_1)} G_\sigma(r^*(\sigma)) \leq -Cn\varepsilon_1.$$

Let g be an entire function of exponential type σ. Let us define $\delta = \delta(\varepsilon)$: $(0,\infty) \to (0, r_0)$ by the equation
$$r := r^*\big(\sigma_0/(1+\varepsilon)\big) = r_0 - \delta.$$

Equivalently, δ is defined by (6.4.6) for $\varepsilon_1 = \varepsilon/(1+\varepsilon)$. Then the left inequality in (3.2.23) of Proposition 3.2.5 shows that for all $\varepsilon \in (0, 0.2C_2)$, we have $\delta \in (0, 0.2)$. Next, for any sequence $\{\delta_n\}_{n=1}^\infty$ with $\lim_{n\to\infty} \delta_n = 0$, satisfying (6.0.15), r satisfies (6.1.15) for large enough n since $r < 0.60$. Using now (6.1.18) and (6.4.9) for $\varepsilon_1 = \varepsilon/(1+\varepsilon)$, $\varepsilon \in (0, 0.2C_3)$, we have that the inequalities

$$\begin{aligned}\|g - P_{n-1}\|_{L_{p,W}[-a_n, a_n]} &\leq J_1 \\ &\leq CC(g,\varepsilon) n^{\gamma_2} (r_0 - \delta)^{-1} \exp\big(G_{\sigma(1+\varepsilon)}(r^*(\sigma_0/(1+\varepsilon)))\big) \\ &\leq CC(g,\varepsilon) n^{\gamma_2} \exp\big(G_{\sigma_0/(1+\varepsilon)}(r^*(\sigma_0/(1+\varepsilon)))\big) \\ &\leq CC(g,\varepsilon) n^{\gamma_2} \exp(-C_4 n\varepsilon)\end{aligned}$$

hold true for any $\sigma \in (0, \sigma_0/(1+\varepsilon)^2]$.

Thus replacing $\sigma_0/(1+\varepsilon)^2$ with $\sigma_0/(1+\varepsilon)$, we have that for any $\sigma \in (0, \sigma_0/(1+\varepsilon)]$ and every $\varepsilon > 0$ small enough, the inequality

(6.4.10) $$E_n\big(g, L_{p,W}[-a_n, a_n]\big) \leq CC(g,\varepsilon) n^\gamma \exp(-C_1 \varepsilon n)$$

is valid.

For some weights the constant σ_0 can be computed explicitly (see Theorem 10.1.2), but for a general W we need to estimate σ_0.

Since the function $\frac{x(1+x^2)}{(1+x)(1-x^2)}$ is increasing on $(0,1)$, we obtain by Propositions 3.2.4(a) and 3.3.8

$$\frac{\sigma_0}{b_n} = \frac{h(r_0)(1+r_0^2)}{b_n(1-r_0^2)} \geq \frac{2Cr_0(1+r_0^2)}{(1+r_0)^2(1-r_0^2)}$$

(6.4.11) $$\geq \frac{2Cr_1(1+r_1^2)}{(1+r_1)^2(1-r_1^2)} > 0.35C,$$

where $C \in (0,1)$ is the constant from Proposition 3.2.4(a).

If $\lim_{x\to\infty} T(x) = \infty$, then by Propositions 3.2.4(b) and 3.3.8, we have for all n large enough

(6.4.12) $$\frac{\sigma_0}{b_n} \geq \frac{2r_1(1+r_1^2)}{(1+r_1)^2(1-r_1^2)}(1+o(1)) > 0.35.$$

Finally, it follows from (6.4.10) and (6.4.11) that the number

(6.4.13) $$\beta = \max\left(1, \limsup_{n\to\infty} b_n/\sigma_0\right)$$

belongs to $[1, (0.35C)^{-1}]$ and for any $\sigma \in (0, b_n/(\beta(1+\varepsilon))]$, (2.2.4) holds. This proves statement (a) of Theorem 2.2.2. Statement (b) immediately follows from (6.4.12). ∎

CHAPTER 7

Proofs of the Limit Theorems

Proofs of Theorems 2.1.1–2.1.4 are given in this chapter. They are based on Theorems 2.2.1, 2.2.2, 2.3.2, and also on some properties of entire functions of exponential type discussed in Chapter 5.

Proofs of relations of the form

$$\lim_{n\to\infty} C_n E_n\big(f(\mu_n \cdot), L_{p,W}(G_n)\big) = A_\sigma(f, L_p), \qquad 0 < p \leq \infty,$$

follow the strategy developed by Bernstein [13, 16], Raitsin [105] and the author [40, 44] for the non-weighted L_p-norms, $1 \leq p \leq \infty$, and generalized in [45, 47] for general normed and quasi-normed spaces.

The proofs of Theorem 2.1.1 and 2.1.2 are divided into two parts. We first prove the inequalities like

$$\limsup_{n\to\infty} C_n E_n\big(f(\mu_n \cdot), L_{p,W}(G_n)\big) \leq A_\sigma(f, L_p),$$

by using estimate (2.2.3) for entire functions of exponential type and then establish the relations like

$$\liminf_{n\to\infty} C_n E_n\big(f(\mu_n \cdot), L_{p,W}(G_n)\big) \geq A_\sigma(f, L_p),$$

by applying polynomial inequalities (2.3.4) and (2.3.5).

The proofs of Theorems 2.1.3 and 2.1.4 are based on estimate (2.2.4) and Theorems 2.1.1 and 2.1.2.

We remark that certain details of the proofs can be omitted if we refer to the general limit theorems established in [45, 47]. Nevertheless, for the convenience of the reader, we give the direct proof of each theorem.

7.1. Proof of Theorem 2.1.1

(a) Let $W \in \mathcal{F}(C^2)$ and let $f \in L_\infty(\mathbb{R})$. To establish relations (2.1.1) and (2.1.2), we shall prove the following inequalities:

(7.1.1) $$\limsup_{n\to\infty} E_n\big(f(\gamma_n \cdot), L_{\infty,W}(I)\big) \leq A_\sigma(f, L_\infty),$$

(7.1.2) $$\liminf_{n\to\infty} E_n\big(f(\gamma_n \cdot), L_{\infty,W}[-a_n, a_n]\big) \geq A_\sigma(f, L_\infty),$$

where γ_n is given in (2.1.3).

We first prove inequality (7.1.1). Let us set $\delta := \delta_1$ and $C := C_1$ in (2.1.3), where δ_1 and C_1 are constants from Theorem 2.2.1. By Proposition 5.2.1, there exists $g \in B_\sigma \cap L_\infty(\mathbb{R})$ such that

$$\|f - g\|_{L_\infty(\mathbb{R})} = A_\sigma(f, L_\infty).$$

Hence

$$\limsup_{n\to\infty} E_n\big(f(\gamma_n\cdot), L_{\infty,W}(I)\big)$$
$$\leq \limsup_{n\to\infty} E_n\big(f(\gamma_n\cdot) - g(\gamma_n\cdot), L_{\infty,W}(I)\big) + \limsup_{n\to\infty} E_n\big(g(\gamma_n\cdot), L_{\infty,W}(I)\big)$$
(7.1.3) $\quad \leq A_\sigma(f, L_\infty) + \limsup_{n\to\infty} E_n\big(g(\gamma_n\cdot), L_{\infty,W}(I)\big).$

Since $g(\gamma_n\cdot) \in B_{b_n(1-C_1 n^{-\delta_1})} \cap L_\infty(\mathbb{R})$, where $\delta_1 \in \big(0, \frac{2(\Lambda-1)}{3(\Lambda+1)}\big)$, we have from Theorem 2.2.1 with $m = 0$, $N = 0$, and $p = \infty$,

(7.1.4) $$\lim_{n\to\infty} E_n\big(g(\gamma_n\cdot), L_{\infty,W}(I)\big) = 0.$$

Thus (7.1.3) and (7.1.4) yield (7.1.1).

Let $\delta > 0$ and $C > 0$ be any constants in (2.1.3). To prove (7.1.2), we consider the polynomials $P_n(x) = \sum_{k=0}^{n} d_{k,n} x^k$, $n = 1, 2, \ldots$, satisfying

(7.1.5) $$\|f(\gamma_n\cdot) - P_n\|_{L_{\infty,W}[-a_n, a_n]} = E_n\big(f(\gamma_n\cdot), L_{\infty,W}[-a_n, a_n]\big).$$

Hence

(7.1.6) $$\|P_n\|_{L_{\infty,W}[-a_n, a_n]} \leq 2\|f(\gamma_n\cdot)\|_{L_{\infty,W}[-a_n, a_n]} \leq 2\|f\|_{L_\infty(\mathbb{R})}.$$

Without loss of generality we may assume that the limit $\lim_{n\to\infty} E\big(f(\gamma_n\cdot), L_{\infty,W}[-a_n, a_n]\big)$ exists.

Next by Theorem 2.3.2(a),

(7.1.7) $$|d_{k,n}| \leq \frac{C\sqrt{k+1}\, b_n^k}{k!} \|P_n\|_{L_{\infty,W}[-a_n, a_n]}.$$

Further setting

$$Q_n(x) := P_n(x/\gamma_n) = \sum_{k=0}^{n} c_{k,n} x^k$$

and using (7.1.6), (7.1.7), we obtain the following estimate for the coefficients of Q_n:

(7.1.8) $$|c_{k,n}| \leq \frac{C\sqrt{k+1}\, \sigma^k}{k!(1 - Cn^{-\delta})^k} \|f\|_{L_\infty(\mathbb{R})}, \qquad k = 0, 1, \ldots, n.$$

Then applying Proposition 5.1.2(a) to a sequence $\{Q_n\}_{n=1}^\infty$, we have that there exist a subsequence $\{n_s\}_{s=1}^\infty$ and a function g_0 such that

(7.1.9) $$\lim_{s\to\infty} Q_{n_s}(z) = g_0(z)$$

uniformly in each interval $[-M, M]$. Note that by (7.1.8), $g_0 \in B_{\sigma(1+\tau)}$ for any $\tau \in (0, 1)$.

Since Q is increasing in $(0, c)$, continuous at zero, $Q(0) = 0$ and $\lim_{n\to\infty} \gamma_n = \infty$ (see (3.2.8)), we have that for any $M > 0$ and every $\varepsilon \in (0, 1)$ there exists $n_1 = n_1(M, \varepsilon)$ such that for $n > n_1$

(7.1.10) $$\min_{x \in [-M, M]} \exp\big(-Q(x/\gamma_n)\big) \geq 1 - \varepsilon.$$

Then for given $M > 0$, $\varepsilon \in (0,1)$, and for large enough n we obtain from (7.1.5) and (7.1.10)

$$E_n\big(f(\gamma_n\cdot), L_{\infty,W}[-a_n, a_n]\big) = \max_{x \in [-a_n, a_n]} |f(\gamma_n x) - P_n(x)| e^{-Q(x)}$$

$$= \max_{y \in [-a_n\gamma_n, a_n\gamma_n]} |f(y) - Q_n(y)| \exp(-Q(y/\gamma_n))$$

$$\geq \max_{y \in [-M, M]} |f(y) - Q_n(y)| \exp(-Q(y/\gamma_n)) \geq (1-\varepsilon) \|f - Q_n\|_{L_\infty[-M,M]}.$$

Hence by (7.1.9),

(7.1.11) $\quad \lim_{s \to \infty} E_{n_s}\big(f(\gamma_{n_s}\cdot), L_{\infty,W}[-a_{n_s}, a_{n_s}]\big) \geq (1-\varepsilon) \|f - g_0\|_{L_\infty[-M,M]}.$

Letting $M \to \infty$ in (7.1.11), we get

$$\lim_{n \to \infty} E_n\big(f(\gamma_n\cdot), L_{\infty,W}[-a_n, a_n]\big) \geq (1-\varepsilon) \|f - g_0\|_{L_\infty(\mathbb{R})}$$

(7.1.12) $\hspace{6cm} \geq (1-\varepsilon) A_{\sigma(1+\tau)}(f, L_\infty).$

Next letting $\varepsilon \to 0$ and then $\tau \to 0$ in (7.1.12), we have

$$\lim_{n \to \infty} E\big(f(\gamma_n\cdot), L_{\infty,W}[-a_n, a_n]\big) \geq A_{\sigma+}(f, L_\infty).$$

Since by Proposition 5.2.3(a), $A_{\sigma+}(f, L_\infty) = A_\sigma(f, L_\infty)$, we arrive at (7.1.2).
Then (7.1.1) and (7.1.2) yield (2.1.1) and (2.1.2).

(b) Let f satisfy inequalities (2.1.4) and (2.1.5) and let $C := C_1$, $\delta := \delta_1$. Then by Proposition 5.2.1, there exists $g_0 \in B_{\sigma_0}$ such that

$$A_{\sigma_0}(f, L_\infty) = \|f - g_0\|_{L_\infty(\mathbb{R})} < \infty.$$

Hence the function $F = f - g_0$ belongs to $L_\infty(\mathbb{R})$ and for $\sigma \geq \sigma_0$

(7.1.13) $\hspace{3cm} A_\sigma(f, L_\infty) = A_\sigma(F, L_\infty).$

Next, for any $x \in \mathbb{R}$,

$$|g_0(x)| \leq \|f - g_0\|_{L_\infty(\mathbb{R})} + A(1+|x|)^N \leq C_2(1+|x|)^N.$$

Since (3.2.5) implies $\gamma_n \leq C_4 n$, we deduce the inequality

$$|g_0(\gamma_n x)| \leq C_3 n^N (1+|x|)^N.$$

Using now Theorem 2.2.1 for $p = \infty$, $m = N$, we get

(7.1.14) $\hspace{3cm} \lim_{n \to \infty} E_n\big(g_0(\gamma_n\cdot), L_{\infty,W}(I)\big) = 0.$

Then applying Theorem 2.1.1(a) to $F \in L_\infty(\mathbb{R})$ and using (7.1.13), (7.1.14), we obtain

$$A_\sigma(f, L_\infty) = A_\sigma(F, L_\infty) = \lim_{n \to \infty} E_n\big(F(\gamma_n\cdot), L_{\infty,W}[-a_n, a_n]\big)$$

$$- \lim_{n \to \infty} E_n\big(g_0(\gamma_n\cdot), L_{\infty,W}[-a_n, a_n]\big)$$

$$\leq \lim_{n \to \infty} E_n\big(f(\gamma_n\cdot), L_{\infty,W}(I)\big)$$

$$= \lim_{n \to \infty} E_n\big(F(\gamma_n\cdot), L_{\infty,W}(I)\big) + \lim_{n \to \infty} E_n\big(g_0(\gamma_n\cdot), L_{\infty,W}(I)\big)$$

$$= A_\sigma(F, L_\infty) = A_\sigma(f, L_\infty).$$

Thus relations (2.1.1) and (2.1.2) hold for functions, satisfying (2.1.4) and (2.1.5).
∎

7.2. Proof of Theorem 2.1.2

(a) It is easy to see that (2.1.6) and (2.1.7) follow from the relations

(7.2.1) $$\limsup_{n\to\infty} \beta_n^{1/p} E_n\big(f(\beta_n\cdot), L_{p,W}(I)\big) \leq A_\sigma(f, L_p),$$

(7.2.2) $$\liminf_{n\to\infty} \beta_n^{1/p} E_n\big(f(\beta_n\cdot), L_{p,W}[-a_n, a_n]\big) \geq A_\sigma(f, L_p),$$

where $p \in (0, \infty)$ and
$$\beta_n := b_n/\sigma.$$

Let $f \in L_p(\mathbb{R})$, $0 < p < \infty$. Then by Proposition 5.2.1, there exists $g \in B_\sigma \cap L_p(\mathbb{R})$ satisfying

(7.2.3) $$\|f - g\|_{L_p(\mathbb{R})} = A_\sigma(f, L_p).$$

Next setting $\beta_{n,\varepsilon} := \beta_n(1 - \varepsilon)$, where $\varepsilon \in (0, 1)$, and taking account of semi-additivity of $E_n^{\mu(p)}(f, L_{p,W}(\mathbb{R}))$ (see (1.3.3)), we have

$$\limsup_{n\to\infty} \beta_{n,\varepsilon}^{\mu(p)/p} E_n^{\mu(p)}\big(f(\beta_{n,\varepsilon}\cdot), L_{p,W}(I)\big)$$
$$\leq \limsup_{n\to\infty} \beta_{n,\varepsilon}^{\mu(p)/p} E_n^{\mu(p)}\big(f(\beta_{n,\varepsilon}\cdot) - g(\beta_{n,\varepsilon}\cdot), L_{p,W}(I)\big)$$
$$+ \limsup_{n\to\infty} \beta_{n,\varepsilon}^{\mu(p)/p} E_n^{\mu(p)}\big(g(\beta_{n,\varepsilon}\cdot), L_{p,W}(I)\big)$$

(7.2.4) $$\leq A_\sigma^{\mu(p)}(f, L_p) + \limsup_{n\to\infty} \beta_{n,\varepsilon}^{\mu(p)/p} E_n^{\mu(p)}\big(g(\beta_{n,\varepsilon}\cdot), L_{p,W}(I)\big).$$

Further since $g(\beta_{n,\varepsilon}\cdot) \in B_{b_n(1-\varepsilon)} \cap L_p(\mathbb{R})$, we have by Propositions 5.1.1(c) and 3.2.2(a)

$$|g(\beta_{n,\varepsilon} x)| \leq C b_n^{1/p} \|g\|_{L_p(\mathbb{R})} \leq C n^{1/p} \|g\|_{L_p(\mathbb{R})}, \qquad x \in \mathbb{R}.$$

Applying Theorem 2.2.1 for $N = 0$ and $m = 1/p$ to the function $g(\beta_{n,\varepsilon}\cdot)$, we obtain

(7.2.5) $$\lim_{n\to\infty} \beta_{n,\varepsilon}^{\mu(p)/p} E_n^{\mu(p)}\big(g(\beta_{n,\varepsilon}\cdot), L_{p,W}(I)\big) = 0.$$

Then (7.2.4) and (7.2.5) imply

(7.2.6) $$\limsup_{n\to\infty} \beta_{n,\varepsilon}^{\mu(p)/p} E_n^{\mu(p)}\big(f(\beta_{n,\varepsilon}\cdot), L_{p,W}(I)\big) \leq A_\sigma^{\mu(p)}(f, L_p).$$

Since (7.2.6) holds for all $\sigma > 0$ and any $\varepsilon \in (0, 1)$, we can replace in (7.2.6) σ with $\sigma(1 - \varepsilon)$ and $\beta_{n,\varepsilon}$ with β_n. Then

(7.2.7) $$\limsup_{n\to\infty} \beta_n^{\mu(p)/p} E_n^{\mu(p)}\big(f(\beta_n\cdot), L_{p,W}(I)\big) \leq A_{\sigma(1-\varepsilon)}^{\mu(p)}(f, L_p).$$

Finally letting $\varepsilon \to 0$ in (7.2.7) and taking account of continuity of $A_\tau(f, L_p)$ in $\tau > 0$ (see Proposition 5.2.3), we arrive at (7.2.1).

To prove (7.2.2), we consider polynomials $P_n(x) = \sum_{k=0}^{n} d_{k,n} x^k$, $n = 1, 2, \ldots$, such that

(7.2.8) $$\|\beta_n^{1/p} f(\beta_n\cdot) - P_n\|_{L_{p,W}[-a_n, a_n]} = E_n\big(\beta_n^{1/p} f(\beta_n\cdot), L_{p,W}[-a_n, a_n]\big).$$

Then

(7.2.9) $$\|P_n\|_{L_{p,W}[-a_n, a_n]} \leq 2^{1/\mu(p)} \beta_n^{1/p} \|f(\beta_n\cdot)\|_{L_p(\mathbb{R})} \leq C \|f\|_{L_p(\mathbb{R})}.$$

Without loss of generality we may assume that the limit

$$\lim_{n\to\infty} \beta_n^{1/p} E_n\big(f(\beta_n\cdot), L_{p,W}[-a_n, a_n]\big)$$

exists. Next using Theorem 2.3.2(b), we obtain that for any $\varepsilon \in (0,1)$,

$$|d_{k,n}| \leq C(\varepsilon) \frac{\sqrt{k+1}\, b_n^{k+1/p}}{k!(1-\varepsilon)^k} \|P_n\|_{L_{p,W}[-a_n,a_n]}, \qquad 0 \leq k \leq n. \tag{7.2.10}$$

Further setting

$$Q_n(x) := \beta_n^{-1/p} P_n(x/\beta_n) = \sum_{k=0}^n c_{k,n} x^k,$$

we have from (7.2.9) and (7.2.10) for $0 \leq k \leq n$, $n = 1, 2, \ldots$,

$$|c_{k,n}| \leq C(\varepsilon) \frac{\sqrt{k+1}\, \sigma^k}{k!(1-\varepsilon)^k} \|P_n\|_{L_{p,W}(I)} \leq C(\varepsilon) \frac{\sqrt{k+1}\, \sigma^k}{k!(1-\varepsilon)} \|f\|_{L_p(\mathbb{R})}. \tag{7.2.11}$$

Hence by Proposition 5.1.2(a), there exist a subsequence $\{n_s\}_{s=1}^\infty$ and a function $g_0 \in B_{\sigma/(1-\varepsilon)}$ such that

$$\lim_{s \to \infty} Q_{n_s}(z) = g_0(z) \tag{7.2.12}$$

uniformly in each interval $[-M, M]$, $M > 0$.

Furthermore, for any $M > 0$ and every $\tau \in (0,1)$ there exists $n_1 = n_1(M,\tau)$ such that for all $n > n_1$ the inequality

$$\min_{x \in [-M,M]} \exp\bigl(-Q(x/\beta_n)\bigr) \geq 1 - \tau \tag{7.2.13}$$

holds. Thus for given $M > 0$, $\tau \in (0,1)$ and for large enough n we have from (7.2.8) and (7.2.13)

$$\beta_n^{1/p} E_n\bigl(f(\beta_n \cdot), L_{p,W}[-a_n, a_n]\bigr) = \|\beta_n^{1/p} f(\beta_n \cdot) - P_n\|_{L_{p,W}[-a_n, a_n]}$$

$$= \left(\int_{-a_n \beta_n}^{a_n \beta_n} |f(y) - Q_n(y)|^p \exp\bigl(-pQ(y/\beta_n)\bigr) dy \right)^{1/p}$$

$$\geq \left(\int_{-M}^{M} |f(y) - Q_n(y)|^p \exp\bigl(-pQ(y/\beta_n)\bigr) dy \right)^{1/p}$$

$$\geq (1-\tau) \|f - Q_n\|_{L_p[-M,M]}.$$

Hence using (7.2.12), we get for any $M > 0$ and every $\tau \in (0,1)$

$$\lim_{s \to \infty} \beta_{n_s}^{1/p} E_{n_s}\bigl(f(\beta_{n_s} \cdot), L_{p,W}[-a_{n_s}, a_{n_s}]\bigr) \geq (1-\tau) \|f - g_0\|_{L_p[-M,M]}. \tag{7.2.14}$$

Letting $M \to \infty$ in (7.2.14), we have

$$\liminf_{n \to \infty} \beta_n^{1/p} E_{n_s}\bigl(f(\beta_n \cdot), L_{p,W}[-a_n, a_n]\bigr) \geq (1-\tau) \|f - g_0\|_{L_p(\mathbb{R})}$$

$$\geq (1-\tau) A_{\sigma/(1-\varepsilon)}(f, L_p). \tag{7.2.15}$$

Finally letting $\tau \to 0$ and then $\varepsilon \to 0$ in (7.2.15) and taking account of continuity of $A_\sigma(f, L_p)$ in $\sigma > 0$ (see Proposition 5.2.3), we arrive at (7.2.2).

(b) Let f satisfy conditions (2.1.4) and (2.1.8). Then by Proposition 5.2.1, there exists $g_0 \in B_{\sigma_0}$ such that

$$A_{\sigma_0}(f, L_p) = \|f - g_0\|_{L_p(\mathbb{R})} < \infty, \qquad 0 < p < \infty.$$

Hence the function $F := f - g_0$ belongs to $L_p(\mathbb{R})$ and for $\sigma \geq \sigma_0$

$$A_\sigma(f, L_p) = A_\sigma(F, L_p). \tag{7.2.16}$$

Next using Proposition 5.2.2, we see that for some $C = C(f, \sigma_0)$,
$$|g_0(x)| \leq C(1+|x|)^N, \qquad x \in \mathbb{R}.$$
Since $g_0(\beta_n \cdot) \in B_{b_n \sigma_0/\sigma}$, we have from (3.2.5)
$$|g_0(\beta_n x)| \leq C n^N (1+|x|)^N.$$
Applying now Theorem 2.2.1 for $m = N$, we get
(7.2.17) $$\lim_{n \to \infty} \beta_n^{1/p} E_n\big(g_0(\beta_n \cdot), L_{p,W}(I)\big) = 0, \qquad \sigma > \sigma_0.$$

The following chain of relations follows from (7.2.16), (7.2.17), and Theorem 2.1.2(a) applied to $F \in L_p(\mathbb{R})$:

$$\begin{aligned}
A_\sigma^{\mu(p)}(f, L_p) &= A_\sigma^{\mu(p)}(F, L_p) \\
&= \lim_{n \to \infty} \beta_n^{\mu(p)/p} E_n^{\mu(p)}\big(F(\beta_n \cdot), L_{p,W}[-a_n, a_n]\big) \\
&\quad - \lim_{n \to \infty} \beta_n^{\mu(p)/p} E_n^{\mu(p)}\big(g_0(\beta_n \cdot), L_{p,W}[-a_n, a_n]\big) \\
&\leq \lim_{n \to \infty} \beta_n^{\mu(p)/p} E_n^{\mu(p)}\big(f(\beta_n \cdot), L_{p,W}[-a_n, a_n]\big) \\
&\leq \lim_{n \to \infty} \beta_n^{\mu(p)/p} E_n^{\mu(p)}\big(f(\beta_n \cdot), L_{p,W}(I)\big) \\
&\leq \lim_{n \to \infty} \beta_n^{\mu(p)/p} E_n^{\mu(p)}\big(F(\beta_n \cdot), L_{p,W}(I)\big) \\
&\quad + \lim_{n \to \infty} \beta_n^{\mu(p)/p} E_n^{\mu(p)}\big(g_0(\beta_n \cdot), L_{p,W}(I)\big) \\
&\leq A_\sigma^{\mu(p)}(F, L_p) = A_\sigma^{\mu(p)}(f, L_p), \qquad \sigma > \sigma_0.
\end{aligned}$$

This yields (2.1.6) and (2.1.7) for functions f of at most polynomial growth, satisfying (2.1.5). ∎

7.3. Proofs of Theorems 2.1.3 and 2.1.4

The proofs are similar to those of Theorems 2.1.1(b) and 2.1.2(b). Let $f \in L_{p,\mathrm{loc}}(\mathbb{R})$, $0 < p \leq \infty$, satisfy condition (2.1.5) for some $\sigma_0 > 0$. Let us set
$$\lambda_{n,p} := \begin{cases} b_n/\sigma, & p \in (0, \infty) \\ b_n(1 - Cn^{-\delta})/\sigma, & p = \infty, \end{cases}$$
where δ and C are the constants from Theorem 2.1.1. Next by Proposition 5.2.1, there exists $g_0 \in L_p(\mathbb{R})$, satisfying
$$A_{\sigma_0}(f, L_p) = \|f - g_0\|_{L_p(\mathbb{R})} < \infty, \qquad 0 < p \leq \infty.$$
Then the function $F := f - g_0$ belongs to $L_p(\mathbb{R})$ and for $\sigma \geq \sigma_0$
(7.3.1) $$A_\sigma(f, L_p) = A_\sigma(F, L_p).$$

If $\beta = \beta(W)$ is the constant from Theorem 2.2.2(a), then the function $g_0(\lambda_{n,p} \cdot) \in B_{b_n \sigma_0/\sigma} \subseteq B_{b_n/(\beta(1+\varepsilon))}$, where $\varepsilon \in (0, \sigma/(\beta \sigma_0))$ is small enough. Thus we obtain from Theorem 2.2.2(a)
(7.3.2) $$\lim_{n \to \infty} \lambda_{n,p}^{1/p} E_n\big(g_0(\lambda_{n,p} \cdot), L_{p,W}[-a_n, a_n]\big) = 0.$$

7.3. PROOFS OF THEOREMS 2.1.3 AND 2.1.4

Next applying Theorems 2.1.1(a) and 2.1.2(a) to F for $p = \infty$ and $p \in (0, \infty)$, respectively, we deduce from (7.3.1) and (7.3.2)

$$\begin{aligned}
A_\sigma^{\mu(p)}(f, L_p) &= A_\sigma^{\mu(p)}(F, L_p) = \lim_{n\to\infty} \lambda_{n,p}^{\mu(p)/p} E_n^{\mu(p)}\big(F(\lambda_{n,p}\cdot), L_{p,W}[-a_n, a_n]\big) \\
&\quad - \lim_{n\to\infty} \lambda_{n,p}^{\mu(p)/p} E_n^{\mu(p)}\big(g_0(\lambda_{n,p}\cdot), L_{p,W}[-a_n, a_n]\big) \\
&\leq \lim_{n\to\infty} \lambda_{n,p}^{\mu(p)/p} E_n^{\mu(p)}\big(f(\lambda_{n,p}\cdot), L_{p,W}[-a_n, a_n]\big) \\
&\leq \lim_{n\to\infty} \lambda_{n,p}^{\mu(p)/p} E_n^{\mu(p)}\big(F(\lambda_{n,p}\cdot), L_{p,W}[-a_n, a_n]\big) \\
&\quad + \lim_{n\to\infty} \lambda_{n,p}^{\mu(p)/p} E_n^{\mu(p)}\big(g_0(\lambda_{n,p}\cdot), L_{p,W}[-a_n, a_n]\big) \\
&= A_\sigma^{\mu(p)}(F, L_p) = A_\sigma^{\mu(p)}(f, L_p), \qquad \sigma > \beta\sigma_0.
\end{aligned}$$

This proves Theorem 2.1.3. Further, if a weight $W \in \mathcal{F}(C^2)$ satisfies the condition $\lim_{x\to\infty} T(x) = \infty$, then by Theorem 2.2.2(b), the constant β belongs to $[1, 3]$. Thus Theorem 2.1.4 follows. ∎

Remarks

(a) The proof of Theorem 2.2.2 shows that we can set

$$\beta(W, n) := \frac{b_n(1 - r_0^2)}{h(r_0)(1 + r_0^2)} \tag{7.3.3}$$

and define β by

$$\beta = \beta(W) = \limsup_{n\to\infty} \beta(W, n). \tag{7.3.4}$$

(b) We conjecture that β defined by (7.3.3) and (7.3.4) is the least constant such that Theorems 2.2.2 and 2.1.3 hold. We recall that the least constant $\gamma = 1.5088\ldots$ for the non-weighted approximation was found by Bernstein [4, 16].

CHAPTER 8

Applications

In this chapter, we consider certain applications of the limit theorems to weighted approximation and orthogonal polynomials.

The asymptotic behavior of the approximation error in $L_{p,W}$, $0 < p \leq \infty$, of functions like $|x|^\lambda \log^k |x|$ is discussed in Theorems 8.1.1 and 8.1.2. In particular, we extend to weighted metrics the celebrated asymptotic of Bernstein [11] for $E_n(f_\lambda, L_\infty[-1,1])$, where $f_\lambda(x) := |x|^\lambda$, $\lambda > 0$. In certain cases, it is possible to compute the corresponding limits explicitly (Corollaries 8.1.3 and 8.1.4). These results strengthen and generalize estimates by Kroó and Szabados [65] established for $\lambda = 1$ and $p = \infty$. Proof of Theorem 2.3.3 is based on such an asymptotic and given in this chapter as well.

Sharp constants of weighted approximation for the non-weighted class $W^r H^\lambda$ are found in Theorem 8.2.1.

A general convergence theorem for a sequence of polynomials which are bounded in a weighted metric is also discussed in this chapter (Theorem 8.3.1). A more precise result on the asymptotic behavior of the sequence of orthonormal polynomials with exponential weights (the Mehler-Heine formula) is given in Theorem 8.3.2.

We remark that the constant b_n plays an important role in all these results.

8.1. Approximation of Individual Functions and Proof of Theorem 2.3.3

The asymptotic relation

(8.1.1) $$\lim_{n \to \infty} n^{\lambda + 1/p} E_n(f_\lambda, L_p[-1,1]) = B_{\lambda,p},$$

where $f_\lambda(x) := |x|^\lambda$ and

(8.1.2) $$B_{\lambda,p} := A_1(f_\lambda, L_p), \qquad 0 < p \leq \infty, \quad \lambda > -1/p,$$

immediately follows from (1.1.5) and (1.1.6), provided that the Bernstein constant

(8.1.3) $$B_{\lambda,p} < \infty.$$

We shall show below (see Theorem 8.1.1(b)) that (8.1.3) holds if $p \in (0, \infty]$ and $\lambda > \max(-1, -1/p)$. This improves some estimates given in [49]. However, the problem of finding necessary and sufficient conditions on p and λ for (8.1.3) to be valid, is still open.

Bernstein [11] proved his celebrated asymptotic (8.1.1) for $p = \infty$ and $\lambda > 0$ in 1938 by a different method. Nevertheless, this formula has triggered his research on the limit theorems like (1.1.3) that were established in 1946. Note that (8.1.1) for $p = \infty$ and $\lambda > 0$ easily follows from (1.1.3) as well (cf. [114, p. 416]).

Using Bernstein's algorithms [7], Varga and Carpenter [119] computed

(8.1.4) $$B_{1,\infty} = 0.2801\ldots.$$

The numerical values of $B_{\lambda,\infty}$ for $\lambda \neq 1$ are unknown, though some estimates are given in [11, 49]. An interesting formula for $B_{\lambda,\infty}$ was recently found by Lubinsky [81]. Totik [117] developed a promising approach to approximation of f_λ on compact subsets of $\mathbb{R}$ (see also Vasiliev [120]).

Nikolskii [99] proved (8.1.1) for $p = 1$, $\lambda > -1$, and found the explicit representation for the Bernstein constant

$$B_{\lambda,1} = \frac{4|\sin(\pi\lambda/2)|}{\pi} \int_{-\infty}^{\infty} |\cos t| \int_0^{\infty} \frac{u^{\lambda+1}}{(t^2+u^2)(e^u+e^{-u})} \, du \, dt.$$

Bernstein [18] noticed that this expression can be reduced to

(8.1.5) $$B_{\lambda,1} = (8/\pi)|\sin(\pi\lambda/2)|\Gamma(\lambda+1)\sum_{k=0}^{\infty}(-1)^k(2k+1)^{-\lambda-2}.$$

Relation (8.1.1) for $p \in (1,\infty)$ and $p \in (0,1)$ was proved by Raitsin [105] and by the author [47], respectively, provided that (8.1.3) holds. Also, Ibragimov [56] and Raitsin [106] independently computed

(8.1.6) $$B_{\lambda,2} = (2/\sqrt{\pi})|\sin(\pi\lambda/2)|\Gamma(\lambda+1)(2\lambda+1)^{-1/2}, \quad \lambda > -1/2.$$

Extensions of (8.1.1) for $p = \infty$, $\lambda > 0$, to λ-homogeneous functions and to the functions $\mu_{\lambda,k}(x) := |x|^\lambda \log^k |x|$, $\lambda > 0$, $k = 1, 2, \ldots$, were obtained in [14, 56, 114, p. 416]. In particular, Bernstein [14] established the following asymptotics as $n \to \infty$:

(8.1.7) $$E_n(\mu_{\lambda,k}, L_\infty[-1,1]) = (1 + o(1))$$
$$\times \begin{cases} B_{\lambda,\infty} n^{-\lambda}(\log n)^k, & \text{if } \lambda > 0 \text{ is not an even integer} \\ kA_1(\mu_{\lambda,1}, L_\infty) n^{-\lambda}(\log n)^{k-1}, & \text{if } \lambda > 0 \text{ is an even integer.} \end{cases}$$

Ahundov [2] found the asymptotic behavior of $E_n(\mu_{\lambda,k}, L_1[-1,1])$.

In the weighted approximation, the first estimates for $E_n(f_1, L_{\infty,W}(\mathbb{R}))$, where W is a weight of the special form, were proved in [10, pp. 139–145]. In particular, if $R_{2m} \in \mathcal{P}_{2m}$ is an even polynomial, satisfying $R_{2m}(0) = 1$, and having only pure imaginary zeros z_k, $1 \leq k \leq 2m$, then

(8.1.8) $$\frac{\sqrt{2}-1}{2}\left(\sum_{k=1}^{2m}\frac{1}{|z_k|}\right)^{-1} < E_m(f_1, L_{\infty, R_{2m}^{-1/2}}(\mathbb{R})) \leq 2\left(\sum_{k=1}^{2m}\frac{1}{|z_k|}\right)^{-1}.$$

Using (8.1.8), it is not difficult to obtain an upper estimate for $E_n(f_1, L_{\infty,W_k}(\mathbb{R}))$, where $W_k(x) = \exp(-|x|^k)$ is the canonical weight with an integer $k \geq 1$. Indeed,

$$R_{2kn}(x) := \prod_{j=1}^{n}\left(1 + \frac{4}{(2j-1)^2}\left(\frac{2x^k}{\pi}\right)^2\right) \leq \prod_{j=1}^{\infty}\left(1 + \frac{4}{(2j-1)^2}\left(\frac{2x^k}{\pi}\right)^2\right)$$

(8.1.9) $$= \frac{e^{2|x|^k} + e^{-2|x|^k}}{2} \leq e^{2|x|^k}.$$

Then we have from (8.1.8) and (8.1.9)

$$(8.1.10) \quad E_{kn}(f_1, L_{\infty, W_k}(\mathbb{R})) \leq E_{kn}(f_1, L_{\infty, R_{2kn}^{-1/2}}(\mathbb{R}))$$

$$\leq \frac{(1/k)(n/4)^{1/k}}{\sum_{j=1}^{n}(2j-1)^{-1/k}} < \begin{cases} (\pi/2)(\log(2n+1))^{-1}, & k = 1 \\ C(n+1)^{1/k-1}, & k > 1. \end{cases}$$

For $k = 1$ these estimates were obtained in [10, p. 145] (see also [29]).

Kroó and Sabados [65] used the similar method to extend (8.1.10) to an even weight $W(x) = e^{-Q(x)}$ on $\mathbb{R}$, satisfying a Freud-type condition

$$0 < \Lambda \leq T(x) \leq \Lambda_1 < \infty \quad \text{for sufficiently large } x.$$

Recall that for our weights, $\Lambda > 1$. It was shown in [65] that the following inequalities hold:

$$(8.1.11) \quad M_n^{-1}(Q) \leq E_n(f_1, L_{\infty, W}(\mathbb{R})) \leq C \left(\int_1^{Q^{(-1)}(n)} \frac{Q(x)}{x^2} dx \right)^{-1},$$

where $Q^{(-1)}$ stands for the inverse function of Q, and $M_n(Q)$ is the least constant on the Markov inequality

$$\|P'\|_{L_{\infty, W}(\mathbb{R})} \leq M \|P\|_{L_{\infty, W}(\mathbb{R})}, \quad P \in \mathcal{P}_n.$$

Moreover, under some additional conditions on Q the following inequality is valid [65, 70]:

$$(8.1.12) \quad M_n^{-1} \geq C \left(\int_1^{Q^{(-1)}(n)} \frac{Q(x)}{x^2} dx \right)^{-1}.$$

It immediately follows from (8.1.11) and (8.1.12) that for such weights

$$E_n(f_1, L_{\infty, W}(\mathbb{R})) \sim \left(\int_1^{Q^{(-1)}(n)} \frac{Q(x)}{x^2} dx \right)^{-1}.$$

In this section, we find the asymptotics for the error of polynomial approximation in $L_{p,W}(I)$ of λ-homogeneous functions f on $\mathbb{R}$ and the functions $f(x) \log^k |x|$. In particular, the asymptotic behavior of $E_n(f_\lambda, L_{p,W}(I))$ will be found.

It is easy to see that any nontrivial λ-homogeneous function f on $\mathbb{R}$ has the form

$$f(x) = f_{\lambda, a, b}(x) := \begin{cases} ax^\lambda, & x > 0 \\ b|x|^\lambda, & x < 0, \end{cases}$$

where $\lambda > 0$ and $a, b \in \mathbb{C}$, $|a| + |b| > 0$. In particular,

$$f_{\lambda, 1, 1}(x) = |x|^\lambda, \quad f_{\lambda, 1, -1}(x) = |x|^\lambda \text{sign } x.$$

Let us set

$$h_{\lambda, k}(x) = h_{\lambda, k, a, b}(x) := f_{\lambda, a, b}(x) \log^k |x|, \quad k = 0, 1, \ldots.$$

In the following results, we extend (8.1.1) and (8.1.7) to approximation with exponential weights and generalize and strengthen relations (8.1.11) and (8.1.12) for $W \in \mathcal{F}(C^2)$:

THEOREM 8.1.1. *Let $W \in \mathcal{F}(C^2)$ and let $0 < p \leq \infty$. Then the following statements hold:*

(a) *For $\lambda > 0$,*

$$\lim_{n \to \infty} b_n^{\lambda+1/p} E_n\big(f_{\lambda,a,b}, L_{p,W}(I)\big) = \lim_{n \to \infty} b_n^{\lambda+1/p} E_n\big(f_{\lambda,a,b}, L_{p,W}[-a_n, a_n]\big)$$
(8.1.13)
$$= A_1(f_{\lambda,a,b}, L_p) < \infty.$$

(b) *For $\lambda > \max(-1, -1/p)$,*

$$\lim_{n \to \infty} b_n^{\lambda+1/p} E_n\big(f_{\lambda,1,1}, L_{p,W}(I)\big) = \lim_{n \to \infty} b_n^{\lambda+1/p} E_n\big(f_{\lambda,1,1}, L_{p,W}[-a_n, a_n]\big)$$
(8.1.14)
$$= B_{\lambda,p} < \infty.$$

THEOREM 8.1.2. *Let $W \in \mathcal{F}(C^2)$ and let $0 < p \leq \infty$. Then the following statements hold:*

(a) *If $f_{\lambda,a,b}$ is a polynomial (i.e., λ is even and $a = b$ or λ is odd and $a = -b$), then for $\lambda > 0$ and $k = 1, 2, \ldots$,*

$$(8.1.15) \quad \lim_{n \to \infty} \frac{b_n^{\lambda+1/p}}{(\log b_n)^{k-1}} E_n\big(h_{\lambda,k}, L_{p,W}(I)\big)$$
$$= \lim_{n \to \infty} \frac{b_n^{\lambda+1/p}}{(\log b_n)^{k-1}} E_n\big(h_{\lambda,k}, L_{p,W}[-a_n, a_n]\big) = k A_1(h_{\lambda,1}, L_p) < \infty.$$

(b) *If $f_{\lambda,a,b}$ is not a polynomial, then for $\lambda > 0$ and $k = 1, 2, \ldots$,*

$$(8.1.16) \quad \lim_{n \to \infty} \frac{b_n^{\lambda+1/p}}{(\log b_n)^k} E_n\big(h_{\lambda,k}, L_{p,W}(I)\big)$$
$$= \lim_{n \to \infty} \frac{b_n^{\lambda+1/p}}{(\log b_n)^k} E_n\big(h_{\lambda,k}, L_{p,W}[-a_n, a_n]\big) = A_1(f_{\lambda,a,b}, L_p) < \infty.$$

(c) *For $\lambda > \max(-1, -1/p)$, $a = b = 1$, and $k = 1, 2, \ldots$, relations (8.1.15) and (8.1.16) hold with $A_1(f_{\lambda,1,1}, L_p) = B_{\lambda,p}$.*

For some λ, p, a, and b, it is possible to compute the limits in Theorems 8.1.1 and 8.1.2 explicitly.

COROLLARY 8.1.3. *If $W \in \mathcal{F}(C^2)$, then for $k = 1, 2, \ldots$, the following statements hold:*

(a) *If λ is not an even number, then*

8.1. APPROXIMATION OF INDIVIDUAL FUNCTIONS AND PROOF OF THEOREM 2.3.3

$$\lim_{n\to\infty} b_n^{\lambda+1/p} E_n\big(f_{\lambda,1,1}, L_{p,W}(I)\big) = \lim_{n\to\infty} b_n^{\lambda+1/p} E_n\big(f_{\lambda,1,1}, L_{p,W}[-a_n, a_n]\big)$$

$$= \lim_{n\to\infty} \frac{b_n^{\lambda+1/p}}{(\log b_n)^k} E_n\big(h_{\lambda,k,1,1}, L_{p,W}(I)\big)$$

$$= \lim_{n\to\infty} \frac{b_n^{\lambda+1/p}}{(\log b_n)^k} E_n\big(h_{\lambda,k,1,1}, L_{p,W}[-a_n, a_n]\big)$$

(8.1.17)
(8.1.18)
(8.1.19)
$$= \begin{cases} (8/\pi)|\sin(\pi\lambda/2)|\Gamma(\lambda+1)\sum_{s=0}^{\infty}(-1)^s(2s+1)^{-\lambda-2}, \\ \quad \text{if } p = 1, \quad \lambda > -1 \\ (2/\sqrt{\pi})|\sin(\pi\lambda/2)|\Gamma(\lambda+1)(2\lambda+1)^{-1/2} \\ \quad \text{if } p = 2, \quad \lambda > -1/2 \\ 0.2801\ldots \\ \quad \text{if } p = \infty, \quad \lambda = 1. \end{cases}$$

(b) *If λ is not an odd number, then*

$$\lim_{n\to\infty} b_n^{\lambda+1/p} E_n\big(f_{\lambda,1,-1}, L_{p,W}(I)\big)$$

$$= \lim_{n\to\infty} b_n^{\lambda+1/p} E_n\big(f_{\lambda,1,-1}, L_{p,W}[-a_n, a_n]\big)$$

$$= \lim_{n\to\infty} \frac{b_n^{\lambda+1/p}}{(\log b_n)^k} E_n\big(h_{\lambda,k,1,-1}, L_{p,W}(I)\big)$$

$$= \lim_{n\to\infty} \frac{b_n^{\lambda+1/p}}{(\log b_n)^k} E_n\big(h_{\lambda,k,1,-1}, L_{p,W}[-a_n, a_n]\big)$$

(8.1.20)
(8.1.21)
$$= \begin{cases} (8/\pi)|\cos(\pi\lambda/2)|\Gamma(\lambda+1)\sum_{s=0}^{\infty}(2s+1)^{-\lambda-2}, & p = 1, \lambda > 0 \\ (2/\sqrt{\pi})|\cos(\pi\lambda/2)|\Gamma(\lambda+1)(2\lambda+1)^{-1/2} & p = 2, \lambda > 0. \end{cases}$$

COROLLARY 8.1.4. *If $W \in \mathcal{F}(C^2)$, then for $k = 1, 2, \ldots$, the following statements hold:*

(a) *If $\lambda > 0$ is an even number, then*

$$\lim_{n\to\infty} \frac{b_n^{\lambda+1/p}}{(\log b_n)^{k-1}} E_n\big(h_{\lambda,k,1,1}, L_{p,W}(I)\big)$$

$$= \lim_{n\to\infty} \frac{b_n^{\lambda+1/p}}{(\log b_n)^{k-1}} E_n\big(h_{\lambda,k,1,1}, L_{p,W}[-a_n, a_n]\big)$$

(8.1.22)
(8.1.23)
$$= \begin{cases} 4\Gamma(\lambda+1)\sum_{s=0}^{\infty}(-1)^s(2s+1)^{-\lambda-2}, & p = 1 \\ \sqrt{\pi}\Gamma(\lambda+1)(2\lambda+1)^{-1/2}, & p = 2. \end{cases}$$

(b) If $\lambda > 0$ is an odd number, then

$$\lim_{n\to\infty} \frac{b_n^{\lambda+1/p}}{(\log b_n)^{k-1}} E_n(h_{\lambda,k,1,-1}, L_{p,W}(I))$$

$$= \lim_{n\to\infty} \frac{b_n^{\lambda+1/p}}{(\log b_n)^{k-1}} E_n(h_{\lambda,k,1,-1}, L_{p,W}[-a_n, a_n])$$

(8.1.24)
(8.1.25)
$$= \begin{cases} 4\Gamma(\lambda+1) \sum_{s=0}^{\infty} (2s+1)^{-\lambda-2}, & p=1 \\ \sqrt{\pi}\Gamma(\lambda+1)(2\lambda+1)^{-1/2}, & p=2. \end{cases}$$

Remarks

(a) For Erdös weights the constant b_n in relations (8.1.13)–(8.1.25) can be replaced by n/a_n (see Proposition 3.2.2(c)).

(b) Proposition 3.2.2(a) shows that for $W \in \mathcal{F}(C^2)$, $b_n \sim n/a_n$. Then by (8.1.19), we have

(8.1.26) $$E_n(f_1, L_{\infty,W}(I)) \sim a_n/n,$$

while for Erdös weights,

(8.1.27) $$E_n(f_1, L_{\infty,W}(I)) = 0.2801\cdots(a_n/n)(1+o(1)), \quad n \to \infty.$$

(c) Note that for Freud weights on $\mathbb{R}$, relation (8.1.26) can be also proved by using the formula

(8.1.28) $$E_n(f_1, L_{\infty,W}(\mathbb{R})) \sim M_n^{-1},$$

since the relation $M_n^{-1} \sim a_n/n$ for Freud weights on $\mathbb{R}$ is established in [72, p. 294]. Inequalities (8.1.11) and (8.1.12) immediately imply (8.1.28) for Freud weights. However, (8.1.28) is invalid for Erdös weights, since in this case $M_n \sim (n/a_n)\sqrt{T(a_n)}$ [72, p. 294], so that (8.1.27) contradicts (8.1.28).

To prove Theorems 8.1.1 and 8.1.2, we need the following lemma in which certain sufficient conditions on f and p for $A_\sigma(f, L_p) < \infty$ are considered:

LEMMA 8.1.5. *Let $f \in L_{1,\text{loc}}(\mathbb{R})$ and let the Fourier transform $\hat{f}$ of the tempered distribution f be a k-differentiable function outside of $(-\varepsilon, \varepsilon)$, $\varepsilon > 0$, with $\hat{f}^{(\ell)} \in L_1(\mathbb{R} \setminus (-\varepsilon, \varepsilon))$ and $\lim_{|x|\to\infty} \hat{f}^{(\ell)}(x) = 0$, $0 \leq \ell \leq k$. Then for $\sigma \geq \varepsilon$ and $p \in (1/k, \infty]$,*

(8.1.29) $$A_\sigma(f, L_p) < \infty.$$

PROOF. We first extend the function $\hat{f}: \mathbb{R} \setminus (-\varepsilon, \varepsilon) \to \mathbb{C}$ to $\mathbb{R}$ by the formula,

$$F_{k,\varepsilon}(x) = \begin{cases} P_{2k+1}(x), & x \in (-\varepsilon, \varepsilon) \\ \hat{f}(x), & x \in \mathbb{R} \setminus (-\varepsilon, \varepsilon), \end{cases}$$

where $P_{2k+1} \in \mathcal{P}_{2k+1}$ is the Hermite interpolation polynomial, satisfying

$$P_{2k+1}^{(\ell)}(\pm\varepsilon) = \hat{f}^{(\ell)}(\pm\varepsilon), \quad 0 \leq \ell \leq k.$$

Then $F_{k,\varepsilon}^{(\ell)} \in L_1(\mathbb{R})$ and

$$\left|\check{F}_{k,\varepsilon}^{(\ell)}(x)\right| = |x^\ell \check{F}_{k,\varepsilon}|, \quad 0 \leq \ell \leq k, \quad x \in \mathbb{R}.$$

Hence

(8.1.30) $\qquad |\check{F}_{k,\varepsilon}(x)| \le C(1+|x|)^{-k}, \qquad k \ge 1, \quad x \in \mathbb{R}.$

Next, $H := \hat{f} - F_{k,\varepsilon}$ is the tempered distribution with the support in $[-\varepsilon, \varepsilon]$, so that by the generalized Paley-Wiener theorem [112, p. 114], the function

$$g_\varepsilon(x) := \check{H}(x) = f(x) - \check{F}_{k,\varepsilon}(x)$$

belongs to B_ε. Moreover by (8.1.30), we have for $p > 1/k$ that $f - g_\varepsilon \in L_p(\mathbb{R})$. Thus (8.1.29) follows. ∎

Proof of Theorem 8.1.1. (a) We first note that

$$f_{\lambda,a,b}(x) = a x_+^\lambda + b x_-^\lambda,$$

where

$$x_+^\lambda := \begin{cases} x^\lambda, & x > 0 \\ 0, & x \le 0, \end{cases} \qquad x_-^\lambda := \begin{cases} |x|^\lambda, & x < 0 \\ 0, & x \ge 0. \end{cases}$$

Then using the well-known formulae for $\widehat{x_+^\lambda}$ and $\widehat{x_-^\lambda}$ [20, 52], we obtain for $\lambda > -1$

$$\hat{f}_{\lambda,a,b}(x) = A_1 x_+^{-\lambda-1} + A_2 x_-^{-\lambda-1},$$

where A_1 and A_2 are constants. Hence the function $f = f_{\lambda,a,b}$ satisfies all conditions of Lemma 8.1.5 for every $\lambda > 0$, $\varepsilon > 0$ and any integer $k \ge 1$. Therefore by Lemma 8.1.5, for any $\sigma_0 > 0$, $\lambda > 0$, and $p \in (0, \infty]$, we have

(8.1.31) $\qquad A_{\sigma_0}(f_{\lambda,a,b}, L_p) < \infty.$

Also, $f_{\lambda,a,b}$ satisfies condition (2.1.4) for $N = \lambda$ and $A = \max(|a|, |b|)$.

Applying Theorem 2.1.1(b) for $p = \infty$ and Theorem 2.1.2(b) for $0 < p < \infty$ to $f = f_{\lambda,a,b}$, $\sigma_0 \in (0,1)$, and $\sigma = 1$, we arrive at (8.1.13).

(b) Using (8.1.31) for $f_{\lambda+1,1,-1}$, we have that for any $\sigma > 0$, $\lambda > -1$ and $p \in (0, \infty]$, there exists an odd function $g_\sigma \in B_\sigma$ such that

$$|x|^{\lambda+1} \operatorname{sign} x - g_\sigma(x) \in L_p(\mathbb{R}).$$

Since $|x|^\lambda \in L_p[-1,1]$ for $\lambda > -1/p$, $p \in (0,\infty]$, and $g_\sigma(x)/x \in B_\sigma$, we get

$$|x|^\lambda - g_\sigma(x)/x \in L_p(\mathbb{R}), \qquad \lambda > \max(-1, -1/p), \quad p \in (0, \infty].$$

Thus, $B_{\lambda,p} < \infty$ for $\lambda > \max(-1, -1/p)$ and $p \in (0, \infty]$.

We cannot apply Theorems 2.1.1(b) and 2.1.2(b) to $f_{\lambda,1,1}$ because $|x|^\lambda$ does not satisfy condition (2.1.4) for $\max(-1, -1/p) < \lambda < 0$. Nevertheless taking account of the fact that $A_{\sigma_0}(f_{\lambda,1,1}, L_p) = \sigma_0^{-\lambda - 1/p} B_{\lambda,p} < \infty$ for any $\sigma_0 > 0$, $\lambda > \max(-1, -1/p)$, and $p \in (0, \infty]$, we can apply Theorem 2.1.3 to $f = f_{\lambda,1,1}$ and $\sigma_0 \in (0, 1/\beta)$. This proves (8.1.14). ∎

Proof of Theorem 8.1.2. We first note that

$$h_{\lambda,s}(x) = a x_+^\lambda (\log x_+)^s + b x_-^\lambda (\log x_-)^s, \qquad s \ge 1.$$

Then assuming that $\lambda > 0$ is not an integer and using the formulae for the Fourier transform of $x_\pm^\lambda (\log x_\pm)^s$ (see [20]), we have

(8.1.32) $\qquad \hat{h}_{\lambda,s}(x) = x_+^{-\lambda-1} \sum_{\ell=0}^{s} A_{\ell,0} (\log x_+)^\ell + x_-^{-\lambda-1} \sum_{\ell=0}^{s} B_{\ell,0} (\log x_-)^\ell,$

where $A_{\ell,0}$, $B_{\ell,0}$, $0 \le \ell \le s$, are constants. It is not difficult to verify by induction that

$$(\hat{h}_{\lambda,s})^{(m)}(x) = x_+^{-\lambda-1} \sum_{\ell=\max(0,s-m)}^{s} A_{\ell,m}(\log x_+)^\ell$$

(8.1.33)
$$+ x_-^{-\lambda-1} \sum_{\ell=\max(0,s-m)}^{s} B_{\ell,m}(\log x_-)^\ell,$$

where $A_{\ell,m}$, $B_{\ell,m}$, $0 \le \ell \le s$, $m = 1, 2, \ldots$, are constants. Hence the function $h_{\lambda,s}$ satisfies all conditions of Lemma 8.1.5 for $\lambda > 0$, $\varepsilon > 0$ and $s = 1, 2, \ldots$. Therefore, if $\lambda > 0$ is not an integer, then for any $\sigma_0 > 0$, $p \in (0, \infty]$ and every integer $s \ge 1$,

(8.1.34)
$$A_{\sigma_0}(h_{\lambda,s}, L_p) < \infty.$$

If $\lambda > 0$ is an integer, then

(8.1.35) $\hat{h}_{\lambda,s}(x) = x_+^{-\lambda-1} \sum_{\ell=0}^{s} A_{\ell,0}(\log x_+)^\ell + x_-^{-\lambda-1} \sum_{\ell=0}^{s} B_{\ell,0}(\log x_-)^\ell + C\delta^{(\lambda)}(x)$

[20]. To get rid of the δ-function in the right-hand side of (8.1.35), we consider $h_{\lambda,s}^*(x) := h_{\lambda,s}(x) - C_1 x^\lambda$. Then for some constant C_1,

$$\widehat{h_{\lambda,x}^*}(x) = x_+^{-\lambda-1} \sum_{\ell=0}^{s} A_{\ell,0}(\log x_+)^\ell + x_-^{-\lambda-1} \sum_{\ell=0}^{s} B_{\ell,0}(\log x_-)^\ell.$$

Next, (8.1.33) holds with $h_{\lambda,s}$ replaced by $h_{\lambda,s}^*$, so we have by Lemma 8.1.5, $A_{\sigma_0}(h_{\lambda,s}, L_p) = A_{\sigma_0}(h_{\lambda,s}^*, L_p) < \infty$. Thus (8.1.34) is valid for all $\lambda > 0$.

Further setting

$$\gamma_{n,p} := \begin{cases} b_n, & 0 < p < \infty \\ b_n(1 - Cn^{-\delta}), & p = \infty \end{cases}$$

we obtain by the substitution $x = y/\gamma_{n,p}$,

$$E_n\big(h_{\lambda,k}, L_{p,W}(I)\big)$$
$$= \gamma_{n,p}^{-1/p} E_n\big(h_{\lambda,k}((1/\gamma_{n,p})\cdot), L_{p,W((1/\gamma_{n,p})\cdot)}(\gamma_{n,p}I)\big)$$
$$= \gamma_{n,p}^{-(\lambda+1/p)} E_n\bigg(\sum_{s=0}^{k}(-1)^{k-s}\binom{k}{s}\log^{k-s}(\gamma_{n,p})\,h_{\lambda,s},$$

(8.1.36)
$$L_{p,W((1/\gamma_{n,p})\cdot)}(\gamma_{n,p}(I))\bigg).$$

(a) Let $f_{\lambda,a,b} \in \mathcal{P}_\lambda$, $\lambda > 0$. Then (8.1.36) implies that for $n \ge \lambda$,

$$E_n\big(h_{\lambda,k}, L_{p,W}(I)\big) = \gamma_{n,p}^{-(\lambda+1/p)} E_n\bigg(\sum_{s=1}^{k}(-1)^{k-s}\binom{k}{s}\log^{k-s}(\gamma_{n,p})\,h_{\lambda,s},$$

(8.1.37)
$$L_{p,W((1/\gamma_{n,p})\cdot)}(\gamma_{n,p}I)\bigg).$$

Note that $h_{\lambda,s}$ satisfies (2.1.4) for $N = \lambda+1$, $A = A(\lambda,s)$ and recall that by (3.2.8), $\lim_{n\to\infty} b_n = \infty$. Then taking account of (8.1.34) and applying Theorem 2.1.1(b) (for

$p = \infty$) and Theorem 2.1.2(b) (for $0 < p < \infty$) to $f = h_{\lambda,s}$, $1 \le s \le k$, $\sigma_0 \in (0,1)$, we obtain

$$\lim_{n \to \infty} (\log b_n)^{-k+1} E_n \left(\binom{k}{s} \log^{k-s} \gamma_{n,p} \, h_{\lambda,s}, L_{p,W((1/\gamma_{n,p})\cdot)}(\gamma_{n,p}(I)) \right)$$

$$= \binom{k}{s} \lim_{n \to \infty} (\log b_n)^{-s+1} \gamma_{n,p}^{1/p} E_n \left(h_{\lambda,s}(\gamma_{n,p}\cdot), L_{p,W}(I) \right)$$

(8.1.38) $$= \binom{k}{s} A_1(h_{\lambda,s}, L_p) \times \begin{cases} 1, & s = 1 \\ 0, & 1 < s \le k. \end{cases}$$

Next using the triangle inequality (1.3.1), we have from (8.1.37) and (8.1.38)

$$\limsup_{n \to \infty} \left(b_n^{\lambda+1/p} (\log b_n)^{1-k} \right)^{\mu(p)} E_n^{\mu(p)} \left(h_{\lambda,k}, L_{p,W}(I) \right)$$

$$\le \lim_{n \to \infty} (\log b_n)^{-(k-1)\mu(p)} E_n^{\mu(p)} \left(\binom{k}{1} \log^{k-1}(\gamma_{n,p}) \, h_{\lambda,1}, \right.$$

$$\left. L_{p,W((1/\gamma_{n,p})\cdot)}(\gamma_{n,p} I) \right)$$

$$+ \sum_{s=2}^{k} \lim_{n \to \infty} (\log b_n)^{-(k-1)\mu(p)} E_n^{\mu(p)} \left(\binom{k}{s} \log^{k-s}(\gamma_{n,p}) \, h_{\lambda,s}, \right.$$

$$\left. L_{p,W((1/\gamma_{n,p})\cdot)}(\gamma_{n,p} I) \right)$$

(8.1.39) $$= k^{\mu(p)} A_1^{\mu(p)}(h_{\lambda,1}, L_p).$$

We prove the lower estimate similarly,

$$\liminf_{n \to \infty} \left(b_n^{\lambda+1/p} (\log b_n)^{1-k} \right)^{\mu(p)} E_n^{\mu(p)} \left(h_{\lambda,k}, L_{p,W}[-a_n, a_n] \right)$$

(8.1.40) $$\ge k^{\mu(p)} A_1^{\mu(p)}(h_{\lambda,1}, L_p).$$

Thus (8.1.39) and (8.1.40) yield (8.1.15).

(b) Let $f_{\lambda,a,b}$ be not a polynomial, $\lambda > 0$. Since $h_{\lambda,s}$ satisfies conditions (2.1.4) and (8.1.34), we can apply Theorem 2.1.1(b) for $p = \infty$ and Theorem 2.1.2(b) for $0 < p < \infty$ to $f = h_{\lambda,s}$, $\sigma_0 \in (0,1)$, $0 \le s \le k$. Then we have

$$\lim_{n \to \infty} (\log b_n)^{-k} E_n \left(\binom{k}{s} \log^{k-s}(\gamma_{n,p}) \, h_{\lambda,s}, L_{p,W((1/\gamma_{n,p})\cdot)}(\gamma_{n,p} I) \right)$$

$$= \binom{k}{s} \lim_{n \to \infty} (\log b_n)^{-s} \gamma_{n,p}^{1/p} E_n \left(h_{\lambda,s}(\gamma_{n,p}\cdot), L_{p,W}(I) \right)$$

(8.1.41) $$= \binom{k}{s} A_1(h_{\lambda,s}, L_p) \times \begin{cases} 1, & s = 0 \\ 0, & 0 < s \le k. \end{cases}$$

Thus (8.1.36) and (8.1.41) imply (8.1.16).

(c) Similarly to the proof of Theorem 8.1.1(b), we can prove that for any $\sigma_0 > 0$, $p \in (0, \infty]$, $\lambda > \max(-1, -1/p)$, and every integer $k \ge 1$, $A_{\sigma_0}(h_{\lambda,s,1,1}, L_p) < \infty$. Since relations (8.1.36)–(8.1.41) hold for $\lambda > \max(-1, -1/p)$, we conclude that (8.1.15) and (8.1.16) are valid for $a = b = 1$, $p \in (0, \infty]$, $\lambda > \max(-1, -1/p)$, and $k = 1, 2, \ldots$. ∎

To prove Corollaries 8.1.3 and 8.1.4, we need the following Sz-Nagy criterion for approximation by entire functions of exponential type in $L_1(\mathbb{R})$ (see [4]):

LEMMA 8.1.6. *Let f be a function from $L_1(\mathbb{R})$. Then the following statements hold:*

(a) *If f is even, $\hat{f}$ is three times differentiable on $\mathbb{R}$, and*

(8.1.42) $$\hat{f}(x) > 0, \quad \frac{d\hat{f}(x)}{dx} \leq 0,$$
$$\frac{d^2\hat{f}(x)}{dx^2} \geq 0, \quad \frac{d^3\hat{f}(x)}{dx^3} \leq 0, \quad x > 1,$$

then

(8.1.43) $$A_1(f, L_1) = 4(2/\pi)^{1/2} \sum_{s=0}^{\infty} (-1)^s (2s+1)^{-1} \hat{f}(2s+1).$$

(b) *If f is odd, $\hat{f}$ is twice continuously differentiable on $\mathbb{R}$, and*

(8.1.44) $$i\hat{f}(x) > 0, \quad i\frac{d\hat{f}(x)}{dx} \leq 0, \quad i\frac{d^2\hat{f}(x)}{dx^2} \geq 0, \quad x > 1,$$

then

(8.1.45) $$A_1(f, L_1) = 4(2/\pi)^{1/2} \sum_{s=0}^{\infty} (2s+1)^{-1} \hat{f}(2s+1).$$

Proof of Corollaries 8.1.3 and 8.1.4. Note first that statement (a) of Corollary 8.1.3 follows from Theorems 8.1.1(b), 8.1.2(c) and relations (8.1.4)–(8.1.6).

Next, Theorems 8.1.1 and 8.1.2 show that all limit relations in Corollaries 8.1.3(b) and 8.1.4 immediately follow from the formulae:

(8.1.46) $A_1(f_{\lambda,1,-1}, L_p)$
$$= (8/\pi)|\cos(\pi\lambda/2)|\Gamma(\lambda+1) \sum_{s=0}^{\infty} (2s+1)^{-\lambda-2}, \quad p=1, \lambda > 0;$$

(8.1.47) $A_1(f_{\lambda,1,-1}, L_p)$
$$= (2/\sqrt{\pi})|\cos(\pi\lambda/2)|\Gamma(\lambda+1)(2\lambda+1)^{-1/2}, \quad p=2, \lambda > 0;$$

$A_1(h_{\lambda,1,1,1}, L_p)$
(8.1.48) $$= 4\Gamma(\lambda+1) \sum_{s=0}^{\infty} (-1)^s (2s+1)^{-\lambda-2}, \quad p=1, \lambda=2,4,\ldots;$$

$A_1(h_{\lambda,1,1,1}, L_p)$
(8.1.49) $$= \sqrt{\pi}\Gamma(\lambda+1)(2\lambda+1)^{-1/2}, \quad p=2, \lambda=2,4,\ldots;$$

$A_1(h_{\lambda,1,1,-1}, L_p)$
(8.1.50) $$= 4\Gamma(\lambda+1) \sum_{s=0}^{\infty} (2s+1)^{-\lambda-2}, \quad p=1, \lambda=1,3,\ldots;$$

$A_1(h_{\lambda,1,1,-1}, L_p)$
(8.1.51) $$= \sqrt{\pi}\Gamma(\lambda+1)(2\lambda+1)^{-1/2}, \quad p=2, \lambda=1,3,\ldots.$$

To prove (8.1.46)–(8.1.51), we use the well-known formulae for the Fourier transforms of the following tempered distributions [20]:

$$
\begin{aligned}
(8.1.52) \quad \hat{f}_{\lambda,1,-1}(x) &= -i(2/\sqrt{2\pi})\cos(\pi\lambda/2) \\
&\quad \times \Gamma(\lambda+1)|x|^{-\lambda-1}\operatorname{sign} x, \quad \lambda > 0, \\
(8.1.53) \quad \widehat{(h_{\lambda,1,1,1} - Q_\lambda)}(x) &= (-1)^{\lambda/2+1}(\sqrt{2\pi}/2) \\
&\quad \times \Gamma(\lambda+1)|x|^{-\lambda-1}, \quad \lambda = 2, 4, \ldots \\
(8.1.54) \quad \widehat{(h_{\lambda,1,1,-1} - Q_\lambda)}(x) &= i(-1)^{(\lambda+1)/2}(\sqrt{2\pi}/2) \\
&\quad \times \Gamma(\lambda+1)|x|^{-\lambda-1}\operatorname{sign} x, \quad \lambda = 1, 3, \ldots
\end{aligned}
$$

where

$$Q_\lambda(x) = \Psi(\lambda+1)x^\lambda, \quad \lambda = 1, 2, \ldots,$$

and $\Psi(z) := \Gamma'(z)/\Gamma(z)$ is the psi-function.

Since for any $\varepsilon \in (0,1)$ the Fourier transforms $\hat{f}$ given by (8.1.52)–(8.1.54) are infinitely differentiable outside of $(-\varepsilon, \varepsilon)$ with $\hat{f}^{(\ell)} \in L_1(\mathbb{R})$, $\ell = 0, 1, \ldots$, we can use the function $g_\varepsilon \in B_\varepsilon$ whose construction is given in the proof of Lemma 8.1.6. Recall that $f - g_\varepsilon = \check{F}_{k,\varepsilon} \in L_p$, $p \in (1/k, \infty]$, and $\widehat{f - g_\varepsilon}$ is three times differentiable on $\mathbb{R}$ if we choose $k \geq 3$.

Further, it is easy to see that the functions $F_1 := f_{\lambda,1,-1} - g_\varepsilon$, $\lambda > 0$, $F_2 := h_{\lambda,1,1,-1} - Q_\lambda - g_\varepsilon$, $\lambda = 1, 3, \ldots$ are odd and the function $F_3 := h_{\lambda,1,1,1} - Q_\lambda - g_\varepsilon$, $\lambda = 2, 4, \ldots$, is even. Since for $x > 1$, $\hat{F}_m$, $m = 1, 2, 3$, coincide with the Fourier transforms given by (8.1.52), (8.1.54), and (8.1.53), respectively, we conclude that F_1 and F_2 satisfy condition (8.1.44) and F_3 satisfies (8.1.42).

Thus using Lemma 8.1.6, we arrive at relations (8.1.46), (8.1.48), and (8.1.50), while (8.1.47), (8.1.49), and (8.1.51) follow from the equality

$$A_1(f, L_2) = \left(\int_{|x|\geq 1} |\hat{f}(x)|^2 \, dx\right)^{1/2}, \quad f \in L_2(\mathbb{R}),$$

and formulae (8.1.52)–(8.1.54).

This completes the proof of the corollaries. ∎

Proof of Theorem 2.3.3. We first note that the inequality

$$\Psi_n := b_n^{-1} \sup_{P\in\mathcal{P}_n} \sup_{z\in\mathbb{C}} |z|^{-1} \log \frac{|P(z)|}{\|P\|_{L_{\infty,W}(I)}} \leq 1$$

is an immediate consequence of Theorem 2.3.1 if $W \in \mathcal{F}_C$, whence it follows that

$$(8.1.55) \qquad \limsup_{n\to\infty} \Psi_n \leq 1, \quad W \in \mathcal{F}_C.$$

Suppose that

$$\liminf_{n\to\infty} \Psi_n \leq a,$$

where $a \in (0,1)$. We may assume without loss of generality that for all $n > n_0$

$$\Psi_n \leq a^* := (1+a)/2 < 1.$$

Then statements (a) and (b) of Theorem 2.3.2 hold with b_n replaced by a^*b_n, $n > n_0$. Next, similarly to the proof of Theorem 2.1.1(a) we can prove that for any

$f \in L_\infty(\mathbb{R})$ and every $\sigma > 0$,

(8.1.56) $$\limsup_{n \to \infty} E_n(f(\gamma_n \cdot), L_{\infty,W}(I)) \geq A_\sigma(f, L_\infty),$$

where $\gamma_n = a^* b_n (1 - Cn^{-\delta})/\sigma$.

Further, let f satisfy (2.1.4) and (2.1.5) for any $\sigma_0 > 0$. Then for $\sigma_0 := \sigma/(2a^*)$ there exists $g_0 \in B_{\sigma_0}$ such that $f - g_0 \in L_\infty(\mathbb{R})$ and $|g_0(\gamma_n x)| \leq Cn^N(1 + |x|)^N$, $x \in \mathbb{R}$. Applying now (8.1.56) to $f - g_0$ and Theorem 2.2.2 to $g_0(\gamma_n \cdot) \in B_{b_n/2}$, we get

$$\begin{aligned} A_\sigma(f, L_\infty) &\leq \limsup_{n \to \infty} E_n((f - g_0)(\gamma_n \cdot), L_{\infty,W}(I)) \\ &\leq \limsup_{n \to \infty} E_n(f(\gamma_n \cdot), L_{\infty,W}(I)) + \lim_{n \to \infty} E_n(g_0(\gamma_n \cdot), L_{\infty,W}(I)) \\ &= \limsup_{n \to \infty} E_n(f(\gamma_n \cdot), L_{\infty,W}(I)). \end{aligned}$$

This shows that (8.1.56) is valid for all functions, satisfying conditions (2.1.4) and (2.1.5) for all $\sigma_0 > 0$. Since the function $f_1(x) = |x|$ satisfies these conditions (see Lemma 8.1.5), we have from (8.1.56) for $\sigma = 1$

$$\limsup_{n \to \infty} b_n E_n(f_1, L_{\infty,W}(I)) \geq B_{1,\infty}/a^*,$$

where $B_{1,\infty}$ is computed in (8.1.4). This contradicts to the following special case of Theorem 8.1.2(b) for $p = \infty$ and $\lambda = 1$:

$$\lim_{n \to \infty} b_n E_n(f_1, L_{\infty,W}(I)) = B_{1,\infty}.$$

Thus

(8.1.57) $$\liminf_{n \to \infty} \Psi_n \geq 1.$$

Finally, (8.1.55) and (8.1.57) yield (2.3.6). ∎

8.2. An Asymptotically Sharp Constant of Weighted Approximation on the Class $W^r H^\lambda[I]$

We define the modulus of continuity of a continuous function $f: I \to \mathbb{R}$ by

$$\omega(f, t) := \sup_{|x-y| \leq t,\ x,y \in I} |f(x) - f(y)|, \qquad t \geq 0.$$

We also consider a modulus of continuity $\omega(t): I \to [0, \infty)$, i.e., a continuous subadditive and non-decreasing function, satisfying the condition $\omega(0) = 0$. The class of all real-valued functions f on I having the rth continuous derivative that satisfies the condition $\omega(f^{(r)}, t) \leq \omega(t)$ for all $t \in [0, \infty)$, is denoted by $W^r H_\omega[I]$, $r = 0, 1, \ldots$. In particular, if $\omega(t) = \omega_\lambda(t) := t^\lambda$, $0 < \lambda \leq 1$, we denote $W^r H^\lambda[I] := W^r H_{\omega_\lambda}[I]$.

Let $W^r H_\omega^*$ be a subset of all 2π-periodic functions from $W^r H_\omega[\mathbb{R}]$ and let $\mathcal{T}_n$ be a class of all trigonometric polynomials with real coefficients of degree at most $n-1$.

For a 2π-periodic function $f \in L_\infty(\mathbb{R})$ and for a class K of continuous functions defined either on $\mathbb{R}$ or on I, we set

$$\begin{aligned} E_n^*(f) &:= \inf_{T_n \in \mathcal{T}_n} \|f - T_n\|, & E_n^*(K) &:= \sup_{f \in K} E_n^*(K), \\ A_\sigma(K) &:= \sup_{f \in K} A_\sigma(f, L_\infty), & E_n(K, L_{\infty,W}(I)) &:= \sup_{f \in K} E_n(f, L_{\infty,W}(I)). \end{aligned}$$

8.2. AN ASYMPTOTICALLY SHARP CONSTANT OF WEIGHTED APPROXIMATION

The problems of finding $E_n^*(K), A_\sigma(K)$, and $E_n(K, L_{\infty,W}(I))$ for various classes K of periodic and non-periodic functions and for the non-weighted L_∞ metric have received much attention since the 1960s, see [4, 25, 62, 114] for references and discussions. In particular, Korneichuk [60–62] proved the relation

(8.2.1) $$E_n^*(W^r H_\omega^*) = M(r, \omega, n), \qquad r = 0, 1, \ldots,$$

where ω is a concave modulus of continuity and

$$M(r, \omega, y) := \begin{cases} (1/2)\omega(\pi/y), & r = 0 \\ (1/2)y^{-r} \int_0^\pi \phi_{r-1}(\pi - x)\omega(x/y)\,dx, & r = 1, 2, \ldots. \end{cases}$$

Here, $\phi_k \in \mathcal{P}_k$ is defined by the recurrence relation

$$\phi_0(x) := 1/2, \quad \phi_k(x) = (1/2) \int_0^{\pi-x} \phi_{k-1}(v)\,dv, \qquad k = 1, 2, \ldots.$$

Note that for $\omega(t) = \omega_\lambda(t)$ we have

$$M_\lambda(r, y) := M(r, \omega_\lambda, y) = y^{-(r+\lambda)} M(r, \omega_\lambda, 1).$$

For $\omega(t) = t$ relation (8.2.1) was independently established by Favard [34] and Akhiezer and Krein [6]. There is an asymptotic version of (8.2.1) for polynomial approximation and for a concave modulus of continuity ω:

(8.2.2) $$\begin{aligned} M(r,\omega,n) - \varepsilon_{n,1} n^{-r}\omega(1/n) &\leq E_n(W^r H_\omega[(-1,1)], L_\infty[-1,1]) \\ &\leq M(r,\omega,n) + \varepsilon_{n,2} n^{-r}\omega(1/n), \end{aligned}$$

where $\varepsilon_{n,i} \geq 0$, $n = 1, 2, \ldots$, and $\lim_{n\to\infty} \varepsilon_{n,i} = 0$, $i = 1, 2$. The upper estimate in (8.2.2) was proved by Temlyakov [62], while the lower one was established by Polovina [104].

The author [51] extended (8.2.1) to the class $W^r H_\omega[\mathbb{R}]$, where ω is concave, and to approximation by entire functions of exponential type:

(8.2.3) $$A_\sigma(W^r H_\omega[\mathbb{R}]) = M(r, \omega, \sigma), \qquad \sigma > 0, \ r = 0, 1, \ldots.$$

For $\omega(t) = t$, (8.2.3) was proved by Krein [63], and for $r = 0$, it was established independently by Dzyadyk [31] and Gromov [55].

It was Bernstein [15, 114] who applied the limit theorems to sharp constants of approximation. In particular, he proved the relation

(8.2.4) $$\lim_{n\to\infty} n^{r+\lambda} E_n(W^r H^\lambda[(-1,1)], L_\infty[-1,1]) = A_1(W^r H^\lambda[\mathbb{R}], L_\infty).$$

Below we use the limit theorems of weighted approximation to prove a weighted version of (8.2.4) and (8.2.2) (for $\omega = \omega_\lambda$).

THEOREM 8.2.1. *If $W \in \mathcal{F}(C^2)$, then for $\lambda \in (0, 1]$ and $r = 0, 1, \ldots$,*

(8.2.5) $$\lim_{n\to\infty} b_n^{r+\lambda} E_n(W^r H^\lambda[I], L_{\infty,W}(I)) = M_\lambda(r, 1).$$

PROOF. Note first that it suffices to prove the following weighted analogue of (8.2.4):

(8.2.6) $$\lim_{n\to\infty} b_n^{r+\alpha} E_n(W^r H^\lambda[I], L_{\infty,W}(I)) = A_1(W^r H^\lambda[\mathbb{R}]),$$

since (8.2.6) and (8.2.3) (for $\omega = \omega_\lambda$) yield (8.2.5).

Let $f \in W^r H^\lambda[\mathbb{R}]$. Then using (8.2.3) for any $\sigma_0 > 0$, we have

(8.2.7) $$A_{\sigma_0}(f, L_\infty) \leq M(r, \omega_\lambda, \sigma_0) = M_\lambda(r, \sigma_0) \leq C\sigma_0^{-r-\lambda} < \infty.$$

Further for $r = 0$,
$$|f(x)| \leq |f(0)| + |x|^\lambda, \qquad x \in \mathbb{R},$$
and for $r \geq 1$,
$$\begin{aligned}|f(x) - T_r(f, x)| &= \frac{|x|^r}{(r-1)!}\left|\int_0^1 (1-y)^{r-1}\big(f^{(r)}(xy) - f^{(r)}(0)\big)\, dy\right| \\ &\leq C|x|^{r+\lambda}, \qquad x \in \mathbb{R},\end{aligned}$$
where $T_r(f, \cdot) \in \mathcal{P}_r$ is the Taylor polynomial for f at 0. Hence there is a constant $A = A(f, r, \lambda)$ such that
$$(8.2.8) \qquad |f(x)| \leq A(1 + |x|)^{r+\lambda}, \qquad x \in \mathbb{R}.$$

Inequalities (8.2.7) and (8.2.8) show that f satisfies conditions (2.1.4) and (2.1.5) of Theorem 2.1.1(b) for $\sigma_0 \in (0, 1)$ and $N = r + \lambda$. Then choosing $\varepsilon_n = Cn^{-\delta}$, where C and δ are the constants from Theorem 2.1.1, we have that the function
$$f_n(x) := \big(b_n(1 - \varepsilon_n)\big)^{-(r+\lambda)} f\big(b_n(1 - \varepsilon_n)x\big)$$
belongs to $W^r H^\lambda[\mathbb{R}]$ so that its restriction to I belongs to $W^r H^\lambda[I]$. Therefore by (2.1.2),
$$\begin{aligned}A_1(f, L_\infty) &= \lim_{n \to \infty} \big(b_n(1 - \varepsilon_n)\big)^{r+\lambda} E_n\big(f_n, L_{\infty, W}(I)\big) \\ &\leq \liminf_{n \to \infty} b_n^{r+\lambda} E_n\big(W^r H^\lambda[I], L_{\infty, W}(I)\big).\end{aligned}$$
This implies the inequality
$$(8.2.9) \qquad A_1\big(W^r H^\lambda[\mathbb{R}]\big) \leq \liminf_{n \to \infty} b_n^{r+\lambda} E_n\big(W^r H^\lambda[I], L_{\infty, W}(I)\big).$$

Let now $f \in W^r H^\lambda[I]$. Then there exists a function $F_r \colon \mathbb{R} \to \mathbb{R}$, satisfying the conditions:

(i) The restriction of $F_r - f$ to I is a polynomial from $\mathcal{P}_{r-1}$, where $\mathcal{P}_{-1} := \{0\}$.
(ii) $F_r \in W^r H^\lambda[\mathbb{R}]$.

Indeed, if $I = \mathbb{R}$, then we set $F_r := f$. If $I = (-c, c)$, with $0 < c < \infty$, then we first extend $f^{(r)}$ to $\mathbb{R}$ by
$$(8.2.10) \qquad F_0^*(x) := \begin{cases} f^{(r)}(-c), & x \in (-\infty, -c) \\ f^{(r)}(x), & x \in [-c, c] \\ f^{(r)}(c), & x \in (c, \infty). \end{cases}$$
It is easy to see that $F_0^* \in W^0 H^\lambda[\mathbb{R}]$. Next, we set $F_0 := F_0^*$ if $r = 0$, and for $r \geq 1$, we set
$$(8.2.11) \qquad F_r(x) := \frac{1}{(r-1)!} \int_0^x (x - t)^{r-1} F_0^*(t)\, dt.$$
Then (i) follows from the relation
$$F_r(x) = f(x) - T_{r-1}(f, x), \qquad x \in [-c, c],$$
and (ii) is an easy consequence of (8.2.10) and (8.2.11).

Further, we have for $r = 0$
$$(8.2.12) \qquad |F_0(x) - F_0(0)| \leq |x|^\lambda, \qquad x \in \mathbb{R},$$

8.2. AN ASYMPTOTICALLY SHARP CONSTANT OF WEIGHTED APPROXIMATION

and for $r \geq 1$,

$$|F_r(x) - T_r(F_r, x)| \leq \frac{|x|^r}{(r-1)!} \int_0^1 (1-y)^{r-1} |F_0^*(xy) - F_0^*(0)| \, dy$$

(8.2.13)
$$\leq C|x|^{r+\lambda}, \quad x \in \mathbb{R}.$$

Next using condition (ii) and Eq. (8.2.3), we get

(8.2.14)
$$A_1(F_r, L_\infty) \leq C,$$

where $C = C(r, \lambda)$.

Let us set $f_r := F_r - T_r(F_r, \cdot)$ and for n large enough, $\sigma = \sigma_n := b_n(1-\varepsilon_n) > 1$, where $\varepsilon_n = Cn^{-\delta}$. Then by (8.2.14) and Proposition 5.2.1, there exists $g_\sigma \in B_\sigma$ such that

(8.2.15)
$$A_\sigma(F_r, L_\infty) = A_\sigma(f_r, L_\infty) = \|f_r - g_\sigma\|_{L_\infty(\mathbb{R})}.$$

It follows from (8.2.12)–(8.2.14) that

$$|g_\sigma(x)| \leq A_\sigma(f_r, L_\infty) + |f_r(x)| \leq A_1(F_r, L_\infty) + |x|^{r+\lambda}$$
$$\leq C(1+|x|)^{r+\lambda}, \quad x \in \mathbb{R},$$

where $C = C(r, \lambda)$.

Applying Theorem 2.2.1 to g_σ, we have

(8.2.16)
$$E_n(g_\sigma, L_{\infty, W}(I)) \leq C n^\gamma \exp(-C_1 n^{\delta_2}),$$

where C, C_1, γ, and $\delta_2 > 0$ are independent of f and n. Then using (8.2.16) and condition (i), we obtain for $n \geq r$

$$E_n(f, L_{\infty, W}(I)) = E_n(F_r, L_{\infty, W}(I)) = E_n(f_r, L_{\infty, W}(I))$$
$$\leq \|f_r - g_\sigma\|_{L_\infty(\mathbb{R})} + E_n(g_\sigma, L_{\infty, W}(I))$$
$$\leq A_\sigma(F_r, L_\infty) + C n^\gamma \exp(-C_1 n^{\delta_2})$$

(8.2.17)
$$= A_1(F_r(\sigma^{-1} \cdot), L_\infty) + C n^\gamma \exp(-C_1 n^{\delta_2}).$$

Since by condition (ii), $\sigma^{r+\lambda} F_r(\sigma^{-1}\cdot) \in W^r H^\lambda[\mathbb{R}]$, we have from (8.2.17) for $n \geq r$

$$(b_n(1-\varepsilon_n))^{r+\lambda} E_n(W^r H^\lambda[I], L_{\infty, W}(I)) \leq A_1(W^r H^\lambda[\mathbb{R}])$$
$$+ C b_n^{r+\lambda}(1-\varepsilon_n)^{r+\lambda} n^\gamma \exp(-C_1 n^{\delta_2}).$$

Hence

(8.2.18)
$$\limsup_{n\to\infty} b_n^{r+\lambda} E_n(W^r H^\lambda[I], L_{\infty, W}(I)) \leq A_1(W^r H^\lambda[\mathbb{R}]).$$

Thus (8.2.9) and (8.2.18) yield (8.2.6). ∎

Some special cases of Theorem 8.2.1 are given below.

COROLLARY 8.2.2. *If $W \in \mathcal{F}(C^2)$, then for $\lambda \in (0,1]$,*

(8.2.19)
$$\lim_{n\to\infty} b_n^\lambda E_n(W^0 H^\lambda[I], L_{\infty, W}(I)) = (1/2)\pi^\lambda,$$
$$\lim_{n\to\infty} b_n^{1+\lambda} E_n(W^1 H^\lambda[I], L_{\infty, W}(I)) = (1/4)\pi^{1+\lambda}/(1+\lambda),$$
$$\lim_{n\to\infty} b_n^{2+\lambda} E_n(W^2 H^\lambda[I], L_{\infty, W}(I)) = (1/8)\pi^{2+\lambda}/(2+\lambda),$$
$$\lim_{n\to\infty} b_n^{r+1} E_n(W^r H^1[I], L_{\infty, W}(I)) = (4/\pi) \sum_{k=0}^{\infty} (-1)^k (2k+1)^{-(r+2)}.$$

Remarks

(a) Relations (8.1.19) and (8.2.19) show that the decay of $E_n(f_1, L_{\infty,W}(I))$ is approximately 5.6 times faster than $E_n(W^0 H^1[I], L_{\infty,W}(I))$.

(b) It follows from (3.2.7) that for Erdös weights, b_n in (8.2.5) can be replaced with n/a_n.

(c) Proposition 3.2.2(a) and Theorem 8.2.1 show that for any $f \in W^r H^\lambda[I]$,

$$E_n(f, L_{\infty,W}(I)) \leq C b_n^{-(r+\lambda)} \leq C_1 (a_n/n)^{r+\lambda},$$

and these estimates cannot be improved on $W^r H^\lambda[I]$. More general results for Freud weights on $\mathbb{R}$ are discussed in [91]. Damelin, Ditzian, Lubinsky, and Totik (see [21, 22, 26, 27, 79]) introduced more sophisticated moduli of smoothness for Freud and Erdös weights on $\mathbb{R}$ and proved the direct and inverse theorems of polynomial approximation in fairly general settings.

8.3. Convergence of Polynomials and a Mehler-Heine Formula for Orthonormal Polynomials

Bernstein [12] noticed that for any sequence of polynomials $P_n \in \mathcal{P}_n$, $n = 1, 2, \ldots$, satisfying $\sup_n \|P_n\|_{L_\infty[-1,1]} < \infty$, there exists a subsequence $\{n_k\}_{k=1}^\infty$ and $g_0 \in B_1$ such that

$$\lim_{k \to \infty} P_{n_k}(z/n_k) = g_0(z)$$

uniformly in any compact subset of $\mathbb{C}$.

The following is an analogue of this result for L_p-metrics with exponential weights.

THEOREM 8.3.1. *Let $W \in \mathcal{F}_C$ and let $P_n \in \mathcal{P}_n$, $n = 1, 2, \ldots$ be a sequence of polynomials, satisfying*

(8.3.1) $$\sup_n \|P_n\|_{L_{p,W}(I)} < \infty, \qquad 0 < p \leq \infty.$$

Then there exist a subsequence $\{n_k\}_{k=1}^\infty$ and a function $g_0 \in B_1$ such that

(8.3.2) $$\lim_{k \to \infty} b_{n_k}^{-1/p} P_{n_k}(z/b_{n_k}) = g_0(z)$$

uniformly in any compact subset of $\mathbb{C}$.

PROOF. Let $P_n(x/b_n) = \sum_{k=0}^n c_{k,n} x^k$, $n = 1, 2, \ldots$, and let (8.3.1) hold.

If $p = \infty$, then by Theorem 2.3.2(a) we have

$$|c_{k,n}| \leq \frac{C\sqrt{k+1}}{k!} \|P_n\|_{L_{\infty,W}(I)}, \qquad 0 \leq k \leq n, \quad n = 1, 2, \ldots.$$

Hence (8.3.2) follows from (8.3.1) and Proposition 5.1.2(a).

If $p \in (0, \infty)$, then by Theorem 2.3.2(b), for every $\varepsilon \in (0, 1)$ there exists a constant $C(\varepsilon)$ such that

$$b_n^{-1/p} |c_{k,n}| \leq \frac{C(\varepsilon)\sqrt{k+1}}{k!(1-\varepsilon)^k} \|P_n\|_{L_{p,W}(I)}, \qquad 0 \leq k \leq n, \quad n = 1, 2, \ldots.$$

Then (8.3.2) with $g_0 \in B_{1/(1-\varepsilon)}$ follows from (8.3.1) and Proposition 5.1.2(a). Since $\bigcap_{\varepsilon \in (0,1)} B_{1/(1-\varepsilon)} = B_1$, we obtain that $g_0 \in B_1$. ∎

8.3. CONVERGENCE OF POLYNOMIALS AND A MEHLER-HEINE FORMULA

We remark that the similar convergence results were used in the proofs of Theorems 2.1.1 and 2.1.2.

For some special sequences of polynomials, there are more precise relations than (8.3.2). For example, the Mehler-Heine formula for the orthonormal Hermite polynomials H_n^* associated with the weight $\exp(-2x^2)$ on $\mathbb{R}$ [113] can be given in the following forms:

$$\lim_{n=2k\to\infty} (-1)^{n/2} n^{1/4} H_n^*\left(z/(2\sqrt{n})\right) = (2/\pi)^{1/2} \cos z, \tag{8.3.3}$$

$$\lim_{n=2k+1\to\infty} (-1)^{(n-1)/2} n^{1/4} H_n^*\left(z/(2\sqrt{n})\right) = (2/\pi)^{1/2} \sin z. \tag{8.3.4}$$

The convergence in these formulae is uniform in any compact subset of $\mathbb{R}$.

Extensions of (8.3.3) and (8.3.4) to orthonormal polynomials

$$p_n(x) = p_n(x, W^2) = \gamma_n x^n + \cdots, \qquad \gamma_n > 0,$$

associated with a weight $W^2(x) = \exp(-2Q(x))$ on I are given below.

THEOREM 8.3.2. *Let $W \in \mathcal{F}(C^2)$. Then there exists a constant $a = a(W) \in (0, \min(\Lambda - 1, 1))$ such that*

$$\lim_{n=2k\to\infty} (-1)^{n/2} a_n^{1/2} p_n(y/b_n) = (2/\pi)^{1/2} \cos y, \tag{8.3.5}$$

$$\lim_{n=2k+1\to\infty} (-1)^{(n-1)/2} a_n^{1/2} p_n(y/b_n) = (2/\pi)^{1/2} \sin y, \tag{8.3.6}$$

uniformly in the interval $[-c_n, c_n]$ with $c_n = o\left(\min\left(b_n, \frac{a}{n^{2+a}}\right)\right)$.

We deduce this result from the following pointwise asymptotic for p_n on I [72, Theorem 15.3]:

LEMMA 8.3.3. *Let $W \in \mathcal{F}(C^2)$. Then there exists $\eta > 0$ such that uniformly for $|x| \leq 1 - n^{-\eta}$, $n \to \infty$, we have*

$$a_n^{1/2} p_n(a_n x) W(a_n x)(1-x^2)^{1/4}$$

$$= (2/\pi)^{1/2} \cos\left(\pi n \int_x^1 \sigma_n^*(t)\, dt + \frac{\arccos x}{2} - \frac{\pi}{4}\right) + O(n^{-\eta}), \tag{8.3.7}$$

where

$$\sigma_n^*(t) := \frac{a_n \sqrt{1-t^2}}{\pi^2 n} \int_{-1}^1 \frac{Q'(a_n v) - Q'(a_n t)}{(v-t)\sqrt{1-v^2}}\, dv$$

is the renormalized density function defined on $[-1, 1]$.

Proof of Theorem 8.3.2. We first establish the following properties of σ_n^*:

(a) $\int_{-1}^1 \sigma_n^*(t)\, dt = 1$.
(b) σ_n^* is an even function on $[-1, 1]$.
(c) $\sigma_n^*(0) = a_n b_n/(\pi n)$.
(d) $\sigma_n^*(t)$ is continuous for $|t| \leq 1/2$.
(e) There exists a constant $a = a(W) \in \min(\Lambda - 1, 1)$ such that

$$\int_0^x \sigma_n^*(t)\, dt = \sigma_n^*(0) x + O(x^{1+a/2}), \qquad x \to 0+,$$

where $O(x^{1+a/2})$ is independent of n.

Property (a) is established in [72, p. 17]. Next, the functions $Q'(a_n v)$ and $tQ'(a_n v) - vQ'(a_n t)$ are odd in $v \in [-1, 1]$, whence it follows that the function

$$\begin{aligned}\sigma_n^*(t) &= \frac{a_n\sqrt{1-t^2}}{\pi^2 n} \int_{-1}^{1} \frac{vQ'(a_n v) - tQ'(a_n t) + tQ'(a_n v) - vQ'(a_n t)}{(v^2 - t^2)\sqrt{1 - v^2}} \\ &= \frac{a_n\sqrt{1-t^2}}{\pi^2 n} \int_{-1}^{1} \frac{vQ'(a_n v) - tQ'(a_n t)}{(v^2 - t^2)\sqrt{1 - v^2}}\end{aligned}$$

is even on $[-1, 1]$. Hence property (b) follows. Further by (3.2.4),

$$\sigma_n^*(0) = \frac{2a_n}{\pi^2 n} \int_0^1 \frac{Q'(a_n v)}{v\sqrt{1-v^2}} dv = \frac{a_n b_n}{\pi n}.$$

This yields property (c).

Since $W \in \mathcal{F}(C^2)$, we have by (1.4.7), that $W \in \mathcal{F}(\mathrm{Lip})$, i.e., $W \in \mathcal{F}$ and there exist constants $a = a(W) \in (0, \min(\Lambda - 1, 1))$ and C, ε_1, satisfying property (a) of Definition 1.4.4. It is known [72, p. 149] that for such a weight,

$$(8.3.8) \quad |\sigma_n^*(u) - \sigma_n^*(v)| \leq \frac{C}{\sqrt{1-|u|}} \left(\frac{|u-v|}{1 - \max(|u|, |v|)} \right)^{a/2}, \quad u, v \in (-1, 1).$$

Hence property (d) follows. Finally using (d), we have

$$\int_0^x \sigma_n^*(t)\, dt = \sigma_n^*(0) x + r_n(x), \quad x > 0,$$

where for some $x_1 \in (0, x)$,

$$r_n(x) = \big(\sigma_n^*(x_1) - \sigma_n^*(0)\big) x.$$

Then using (8.3.8), we arrive at property (e) with the error term, satisfying $|r_n(x)| \leq Cx^{1+a/2}$, where C is independent of n and x.

Next we need the following elementary asymptotics:

$$(8.3.9) \quad \frac{\arccos x}{2} = \frac{\pi}{4} - \frac{x}{2} + O(x^3), \quad x \to 0+,$$
$$(8.3.10) \quad (1-x^2)^{-1/4} = 1 + O(x^2), \quad x \to 0+$$
$$(8.3.11) \quad \exp(Q(a_n x)) = 1 + O(Q(a_n x)) = 1 + O(a_n x), \quad a_n x \to 0+.$$

Since properties (a) and (b) imply

$$(8.3.12) \quad \int_x^1 \sigma_n^*(t)\, dt = 1/2 - \int_0^x \sigma_n^*(t)\, dt,$$

we obtain from (8.3.7), (8.3.9)–(8.3.12) and properties (c), (e) as $a_n x \to 0+$,

$$\begin{aligned}a_n^{1/2} p_n(a_n x) &= (2/\pi)^{1/2}\big(1 + O(a_n x)\big)\big(1 + O(x^2)\big) \\ &\quad \times \cos\left(\pi n/2 - \pi n \int_0^x \sigma_n^*(t)\, dt - x/2 + O(x^3)\right) + O(n^{-\eta}) \\ &= (2/\pi)^{1/2}\big(1 + O(a_n x)\big) \cos\big(\pi n/2 - a_n b_n x \\ (8.3.13) &\quad - x/2 + O(nx^{1+a/2})\big) + O(n^{-\eta}).\end{aligned}$$

Let us set $y := a_n b_n x$, where $|y| \leq c_n = o\big(\min(b_n, n^{\frac{a}{2+a}})\big)$. Then $a_n|x| \leq c_n/b_n = o(1)$, $|x| = o(a_n^{-1})$, and $n|x|^{1+a/2} \leq Cn(c_n/n)^{1+a/2} = o(1)$, by (3.2.5).

Thus we have from (8.3.13)
$$a_n^{1/2} p_n(y/b_n) = (2/\pi)^{1/2}(1 + o(1)) \cos(\pi n/2 - y) + o(1),$$
as $n \to \infty$, where both error terms are independent of $y \in [-c_n, c_n]$. Therefore Theorem 8.3.2 follows. ∎

CHAPTER 9

Multidimensional Limit Theorems of Polynomial Approximation with Exponential Weights

In this chapter, we extend the limit theorems to multidimensional approximation with certain exponential weights and discuss their applications to approximation of λ-homogeneous functions in L_p.

Throughout the chapter, $\mathbb{R}^m$ is the m-dimensional Euclidean space; $\mathbb{C}^m = \mathbb{R}^m + i\mathbb{R}^m$ the m-dimensional complex space; $x = (x_1, \ldots, x_m)$, $y = (y_1, \ldots, y_m)$ points in $\mathbb{R}^m$; $(x, y) := \sum_{i=1}^{m} x_i y_i$; $|x| := (x, x)^{1/2}$.

9.1. Multidimensional Limit Theorems with Exponential Weights

Multidimensional versions of (1.1.5) and (1.1.6) have been obtained in the 1980s and 1990s. In particular, for a continuous or measurable function f of a polynomial growth on $\mathbb{R}^m$ the following relations hold [40, 44]:

$$\lim_{n \to \infty} \inf_{P \in \mathcal{P}_{n,m}} \|f(n(1 - Cn^{-\delta})\cdot) - P\|_{L_\infty(V^*)}$$
(9.1.1)
$$= \inf_{g \in B_V} \|f - g\|_{L_\infty(\mathbb{R}^m)}$$

$$\lim_{n \to \infty} \inf_{P \in \mathcal{P}_{n,m}} \|f(n\cdot) - P\|_{L_p(V^*)}$$
(9.1.2)
$$= \inf_{g \in B_V} \|f - g\|_{L_p(\mathbb{R}^m)}, \quad p \in (0, \infty),$$

where $\delta \in (0, 2/3)$, $\mathcal{P}_{n,m}$ is the class of algebraic polynomials in m variables of total degree at most n, V a convex centrally-symmetric body in $\mathbb{R}^m$, V^* the polar of V, and B_V is the class of entire functions of exponential type with the spectra in V. Relations (9.1.1) and (9.1.2) can be extended to any measurable function on $\mathbb{R}^m$ [40, 44].

To discuss a different multidimensional version of (1.1.5), obtained in [43], and to describe the main results of the chapter, we need some additional notation.

Let $\mathcal{P}_{\mu n,m}$, where $\mu = (\mu_1, \ldots, \mu_m)$ with $\mu_k > 0$, $1 \leq k \leq m$, be the class of all algebraic polynomials of m variables and of degree at most $\mu_k n$ with respect to the kth variable, $1 \leq k \leq m$. Next, let $B_{\sigma,m}$, where $\sigma = (\sigma_1, \ldots, \sigma_m)$ with $\sigma_k > 0$, $1 \leq k \leq m$, be the class of all entire functions g of m variables, satisfying the condition: for every $\varepsilon > 0$ there exists $C(g, \varepsilon)$ such that for all $z = (z_1, \ldots, z_m) \in \mathbb{C}^m$,

$$|g(z)| \leq C(g, \varepsilon) \exp\left((1+\varepsilon) \sum_{k=1}^{m} \sigma_k |z_k|\right).$$

Further, we define the operator $A_\alpha \colon \mathbb{R}^m \to \mathbb{R}^m$ by

$$A_\alpha x := (\alpha_1 x_1, \ldots, \alpha_m x_m), \quad \alpha = (\alpha_1, \ldots, \alpha_m), \quad x \in \mathbb{R}^m.$$

We also need the sets $\mathcal{I}_m := \prod_{k=1}^{m} I_k$, where $I_k = (-c_k, c_k)$, $1 \leq k \leq m$, and

$$\mathcal{A}_{n,m} = \mathcal{A}_{n,m}(W) := \prod_{k=1}^{m} \left[-a_{[\mu_k n]}(W_k), a_{[\mu_k n]}(W_k)\right].$$

Here $[t]$ stands for the integral part of t. We shall use the constants $\delta(W)$, $\delta_1(W)$, and $\delta_2(W)$ from Theorems 2.1.1 and 2.2.1 as well.

Let us define the approximation errors by

$$E_{\mu n, m}(f, L_{p,W}(\Omega)) := \inf_{P \in \mathcal{P}_{\mu n, m}} \|f - p\|_{L_{p,W}(\Omega)},$$

$$A_{\sigma, m}(f, L_p) := \inf_{g \in B_{\sigma, m}} \|f - g\|_{L_p(\mathbb{R}^m)},$$

where $\Omega \subseteq \mathbb{R}^m$, W is a weight on Ω and $\|f\|_{L_{p,W}(\Omega)} := \|fW\|_{L_p(\Omega)}$.

It was proved in [43] that for any $f \in L_\infty(\mathbb{R}^m)$

(9.1.3) $$\lim_{n \to \infty} E_{\mu n, m}\left(f(\mathcal{A}_{n(1-Cn^{-\delta})\alpha} \cdot), L_{\infty, 1}(K_m)\right) = A_{\sigma, m}(f, L_\infty),$$

where $\delta \in (0, 2/3)$, $K_m = [-1, 1]^m$ is the unit cube in $\mathbb{R}^m$, and $\alpha := (\mu_1/\sigma_1, \ldots, \mu_m/\sigma_m)$.

Here, we establish weighted analogues of (9.1.3) in L_p for weights of the form

$$W(x) = \exp\left(-\sum_{k=1}^{m} Q_k(x_k)\right) = \prod_{k=1}^{m} W_k(x_k) := \prod_{k=1}^{m} \exp(-Q_k(x_k)).$$

Note that polynomial approximation with exponential weights of this form was discussed in [92].

THEOREM 9.1.1. *Let $W_k = \exp(-Q_k) \in \mathcal{F}(C^2)$ be a weight on $I_k = (-c_k, c_k)$, $1 \leq k \leq m$. Then the following statements hold:*

(a) *For any continuous function $f \in L_\infty(\mathbb{R}^m)$ and every $\sigma = (\sigma_1, \ldots, \sigma_m)$ with positive components, we have*

(9.1.4) $$\lim_{n \to \infty} E_{\mu n, m}\left(f(\mathcal{A}_{\alpha(n)} \cdot), L_{\infty, W}(\mathcal{I}_m)\right)$$
$$= \lim_{n \to \infty} E_{\mu n, m}\left(f(\mathcal{A}_{\alpha(n)} \cdot), L_{\infty, W}(\mathcal{A}_{n,m})\right) = A_{\sigma, m}(f, L_\infty),$$

where

$$\alpha(n) := \left(\frac{b_{[\mu_1 n]}(Q_1)(1 - Cn^{-\delta(W_1)})}{\sigma_1}, \ldots, \frac{b_{[\mu_m n]}(Q_m)(1 - Cn^{-\delta(W_m)})}{\sigma_m}\right).$$

(b) *Let a continuous function f satisfy the conditions*

(9.1.5) $$|f(x)| \leq A(1 + |x|)^N, \quad x \in \mathbb{R}^m,$$

and

(9.1.6) $$A_{\sigma_0, m}(f, L_\infty) < \infty$$

for some $\sigma_0 = (\sigma_{0,1}, \ldots, \sigma_{0,m})$ with positive components. Then (9.1.4) holds for all σ with $\sigma_k \geq \sigma_{0,k}$, $1 \leq k \leq m$.

The following L_p-version of Theorem 9.1.1 is valid for $p \in (0, \infty)$, but for simplicity we assume that $p \geq 1$.

9.1. MULTIDIMENSIONAL LIMIT THEOREMS WITH EXPONENTIAL WEIGHTS

THEOREM 9.1.2. *Let $W_k = \exp(-Q_k) \in \mathcal{F}(C^2)$ be a weight on $I_k = (-c_k, c_k)$, $1 \leq k \leq m$. Then the following statements hold:*

(a) *For any $f \in L_p(\mathbb{R}^m)$, $1 \leq p < \infty$, and every σ with positive components, we have*

$$\lim_{n \to \infty} \prod_{k=1}^{m} \left(b_{[\mu_k n]}/\sigma_k\right)^{1/p} E_{\mu n,m}\left(f(A_{\alpha^*(n)}\cdot), L_{p,W}(\mathcal{I}_m)\right)$$

$$= \lim_{n \to \infty} \prod_{k=1}^{m} \left(b_{[\mu_k n]}/\sigma_k\right)^{1/p} E_{\mu n,m}\left(f(A_{\alpha^*(n)}\cdot), L_{p,W}(\mathcal{A}_{n,m})\right)$$

(9.1.7) $$= A_{\sigma,m}(f, L_p),$$

where

$$\alpha^*(n) := \left(b_{[\mu_1 n]}(Q_1)/\sigma_1, \ldots, b_{[\mu_m n]}(Q_m)/\sigma_m\right).$$

(b) *If a measurable f satisfies (9.1.5) and*

(9.1.8) $$A_{\sigma_0, m}(f, L_p) < \infty, \qquad 1 \leq p < \infty,$$

for some σ_0 with positive components, then relations (9.1.7) hold for all σ with $\sigma_k \geq \sigma_{0,k}$, $1 \leq k \leq m$.

To prove these theorems, we need multidimensional versions of Theorems 2.2.1 and 2.3.2.

THEOREM 9.1.3. *Let $W_k = \exp(-Q_k) \in \mathcal{F}(C^2)$ be a weight on I_k, $1 \leq k \leq m$. Then for every $\sigma = (\sigma_1, \ldots, \sigma_m)$ with $\sigma_k \in \left(0, b_{[\mu_k n]}(1 - C_1 n^{-\delta_1(W_k)})\right]$, $1 \leq k \leq m$, and for any $g \in B_{\sigma,m}$, satisfying the inequality*

(9.1.9) $$|g(x)| \leq An^m(1+|x|)^N, \qquad x \in \mathbb{R}^m,$$

the following estimate holds ($1 \leq p \leq \infty$):

(9.1.10) $$E_{\mu n, m}\left(g, L_{p,W}(\mathcal{I}_m)\right) \leq CAn^\gamma \exp\left(-C_2 n^{C_3 \min_{1 \leq k \leq m} \delta_2(W_k)}\right).$$

THEOREM 9.1.4. *If $W_k \in \mathcal{F}_C$ is a weight on I_k, $1 \leq k \leq m$, then for any polynomial from $\mathcal{P}_{\mu n,m}$ of the form*

(9.1.11) $$P(x) = P(x_1, \ldots, x_m) = \sum_{0 \leq k_1 \leq \mu_1 n} \cdots \sum_{0 \leq k_m \leq \mu_m n} c_{k_1, \ldots, k_m} x_1^{k_1} \cdots x_m^{k_m}$$

the following inequalities hold:

(9.1.12) $$|c_{k_1, \ldots, k_m}| \leq C \prod_{i=1}^{m} \frac{\sqrt{k_i + 1} b_{[\mu_i n]}^{k_i}(Q_i)}{k_i!} \|P\|_{L_\infty, W(\mathcal{A}_{n,m})},$$

$$|c_{k_1, \ldots, k_m}| \leq C(\varepsilon) \prod_{i=1}^{m} \frac{\sqrt{k_i + 1} b_{[\mu_i n]}^{k_i + 1/p}(Q_i)}{k_i!(1-\varepsilon)^{k_i}} \|P\|_{L_p, W(\mathcal{A}_{n,m})},$$

(9.1.13) $$0 < p < \infty, \quad \varepsilon \in (0,1).$$

We remark that the estimates of the coefficients of polynomials $P \in \mathcal{P}_{\mu n, m}$ in the non-weighted L_∞-metric were given by Bernstein [17].

The proofs of Theorems 9.1.1–9.1.4 are discussed in Sections 9.2 and 9.3.

9.2. Proof of Theorem 9.1.3

We first prove (9.1.10) for

$$g(x) = g_t(x) := \exp\left(i \sum_{k=1}^{m} t_k x_k\right),$$

where $t_k \in \left(0, b_{[\mu_k n]}(1 - C_1 n^{-\delta_1(W_k)})\right)$, $1 \leq k \leq m$, and $i = \sqrt{-1}$. Let $P_{n,t_k} \in \mathcal{P}_{[\mu_k n]-1}$ be the Lagrange interpolation polynomial to $\exp(it_k x_k)$ at the zeros $\{x_{j,[\mu_k n]}^{(q)}\}_{j=1}^{[\mu_k n]}$ of the extremal polynomial $T_{[\mu_k n],q}$, $1 \leq k \leq n$. Then the coefficients of $P_{n,t_k}(x_k)$ are linear combinations of $\exp(it_k x_{j,[\mu_k n]}^{(q)})$, $1 \leq j \leq [\mu_k n]$, so that $P_{n,t_k}(x_k)$ is continuous in t_k for each x_k, $1 \leq k \leq m$.

Next, the proof of Theorem 2.2.1 (see (6.3.14)) shows that

$$\|\exp(it_k \cdot) - P_{n,t_k}\|_{L_{p,W_k}(I_k)} \leq C n^\gamma \exp\left(-C_1 n^{\delta_2(W_k)}\right), \quad 1 \leq k \leq m.$$

Hence we have

$$\left(\int_{\mathcal{I}_m} \left(\left|\prod_{k=1}^{m} \exp(it_k x_k) - \prod_{k=1}^{m} P_{n,t_k}(x_k)\right| W(x)\right)^p dx\right)^{1/p}$$

$$= \left(\int_{\mathcal{I}_m} \left|\sum_{k=1}^{m} \left(\prod_{j=k+1}^{m} \exp(it_j x_j) \prod_{j=1}^{k-1} P_{n,t_j}(x_j)\right. \right.\right.$$

$$\left.\left.\left. \times \left(\exp(it_k x_k) - P_{n,t_k}(x_k)\right)\right)\right|^p W(x)\, dx\right)^{1/p}$$

$$\leq \sum_{k=1}^{m} \prod_{j=1}^{k-1} \|P_{n,t_k}\|_{L_{p,W_j}(I_j)} \prod_{j=k+1}^{m} \|\exp(it_j \cdot)\|_{L_{p,W_j}(I_j)}$$

$$\times \|\exp(it_k \cdot) - P_{n,t_k}\|_{L_{p,W_k}(I_k)}$$

(9.2.1)
$$\leq C n^\gamma \exp\left(-C_1 n^{\min_{1\leq k\leq m} \delta_2(W_k)}\right),$$

where C and γ are independent of n and $t = (t_1, \ldots, t_m)$.

Let now $g \in B_{\sigma,m} \cap L_2(\mathbb{R}^m)$, where

$$\sigma_k \in \left(0, b_{[\mu_k n]}(Q_k)(1 - C n^{-\delta_1(W_k)})\right], \quad 1 \leq k \leq m.$$

Then using the Paley-Wiener representation for g [112, Theorem 4.9], we have

(9.2.2)
$$g(x) = \int_{\prod_\sigma} g_t(x) \varphi(t)\, dt,$$

where $\prod_\sigma := \{t \in \mathbb{R}^m : |t_k| \leq \sigma_k,\ 1 \leq k \leq m\}$ is the parallelepiped in $\mathbb{R}^m$ and $\varphi \in L_2(\mathbb{R}^m)$.

Next, the polynomial $P_{n,t}(x) = \prod_{k=1}^{m} P_{n,t_k}(x_k)$ is a continuous function in $t = (t_1, \ldots, t_m)$ for each x. Hence the function

$$Q_n(x) = \int_{\prod_\sigma} P_{n,t}(x) \varphi(t)\, dt$$

belongs to $\mathcal{P}_{\mu n, m}$ and we obtain from (9.2.1)

$$E_{\mu n,m}\bigl(g, L_{p,W}(\mathcal{I}_m)\bigr) \leq \|g - Q_n\|_{L_{p,W}(\mathcal{I}_m)}$$

$$\leq \int_{\Pi_\sigma} \|g_t - P_{n,t}\|_{L_{p,W}(\mathcal{I}_m)} \varphi(t)\, dt$$

$$\leq \left(\prod_{k=1}^{m} \sigma_k\right)^{1/2} \|\varphi\|_{L_2(\mathbb{R}^m)} \max_{t \in \Pi_\sigma} \|g_t - P_{n,t}\|_{L_{p,W}(\mathcal{I}_m)}$$

(9.2.3) $$\leq C\|g\|_{L_2(\mathbb{R}^m)} n^{\gamma_1} \exp\left(-C_1 n^{\min_{1\leq k\leq m} \delta_2(W_k)}\right).$$

This proves (9.1.10) for $g \in B_{\sigma,m} \cap L_2(\mathbb{R}^m)$.

Finally, we assume that a function $g \in B_{\sigma,m}$ satisfies (9.1.9). Without loss of generality we can assume that $\sigma_k = b_{[\mu_k n]}(Q_k)\bigl(1 - Cn^{-\delta_1(W_k)}\bigr)$, $1 \leq k \leq m$, and the number $N \geq 0$ in (9.1.9) is an even integer. Then for every $\varepsilon \in (0, 1/(N+2))$ the function

$$g_1(x) := g\bigl((1 - \varepsilon(N+2))x\bigr) \prod_{k=1}^{m} \left(\frac{\sin(\varepsilon \sigma_k x_k)}{\varepsilon \sigma_k x_k}\right)^{N+2}$$

belongs to $B_{\sigma,m} \cap L_2(\mathbb{R}^m)$ and

$$\|g_1\|_{L_2(\mathbb{R}^m)} \leq CAn^m \varepsilon^{-m/2} \prod_{k=1}^{m} \sigma_k^{-1/2} \leq CAn^{m_1} \varepsilon^{-m/2}.$$

Hence (9.2.3) implies

(9.2.4) $$E_{\mu n,m}\bigl(g_1, L_{p,W}(\mathcal{I}_m)\bigr) \leq CAn^{\gamma_2} \varepsilon^{-m/2} \exp\left(-C_1 n^{\min_{1\leq k\leq m} \delta_2(W_k)}\right).$$

Further we have

(9.2.5) $$\|g - g_1\|_{L_{p,W}(\mathcal{I}_m)} \leq \|g - g\bigl((1 - \varepsilon(N+2))\cdot\bigr)\|_{L_{p,W}(\mathcal{I}_m)}$$
$$+ \left(\int_{\mathcal{I}_m} \left(\left(1 - \prod_{k=1}^{m} \left(\frac{\sin(\varepsilon \sigma_k x_k)}{\varepsilon \sigma_k x_k}\right)^{N+2}\right)|g(x)|W(x)\right)^p dx\right)^{1/p} = J_1 + J_2.$$

To estimate J_2, we need the following elementary inequalities:

(9.2.6) $$1 - \left|\frac{\sin\tau}{\tau}\right| \leq \frac{\tau^2}{6}, \qquad \tau \in \mathbb{R},$$

(9.2.7) $$1 - \prod_{i=1}^{m} a_i \leq m - \sum_{i=1}^{m} a_i, \qquad a_i \in [0,1], \ 1 \leq i \leq m.$$

Note that (9.2.7) is an immediate consequence of the recurrence relation

$$s_{k+1} = s_k + (1 - a_{k+1})\left(1 - \prod_{i=1}^{k} a_i\right),$$

where

$$s_k := k - 1 - \sum_{i=1}^{k} a_i + \prod_{i=1}^{k} a_i, \qquad k = 1, 2, \ldots.$$

Then we have from (9.2.6) and (9.2.7)

$$J_2 \leq C\left(\int_{\mathcal{I}_m}\left(\left(1-\prod_{k=1}^m\left|\frac{\sin(\varepsilon\sigma_k x_k)}{\varepsilon\sigma_k x_k}\right|\right)|g(x)|W(x)\right)^p dx\right)^{1/p}$$

$$\leq C\left(\int_{\mathcal{I}_m}\left(\sum_{k=1}^m\left(1-\left|\frac{\sin(\varepsilon\sigma_k x_k)}{\varepsilon\sigma_k x_k}\right|\right)|g(x)|W(x)\right)^p dx\right)^{1/p}$$

(9.2.8) $$\leq C\varepsilon^2 \max_{1\leq k\leq m}\sigma_k^2\left(\int_{\mathcal{I}_m}(|x|^2|g(x)|W(x))^p dx\right)^{1/p}.$$

Using the estimate $\max_{1\leq k\leq m}\sigma_k \leq \max_{1\leq k\leq m} b_{[\mu_k n]} \leq Cn$ and taking account of (9.1.9), we obtain from (9.2.8)

(9.2.9) $$J_2 \leq CAn^{\gamma_3}\varepsilon^2.$$

To estimate J_1, we need the following fact established by Bernstein [12] (see also [10]): if a function $\varphi \in B_\sigma$ of a single variable satisfies the condition

(9.2.10) $$|\varphi(\tau)| \leq A(a^2+\tau^2)^{N/2}, \qquad \tau \in \mathbb{R}, \quad a \geq 1$$

where $N \geq 0$ is an integer, then

$$|\varphi'(\tau)| \leq A\left|\left((\tau+ia)^N e^{-i\sigma\tau}\right)'\right|.$$

Since for $a \geq 1$

$$\begin{aligned}\left|\left((\tau+ia)^N e^{-i\sigma\tau}\right)'\right| &= (a^2+\tau^2)^{(N-1)/2}\left((N+\sigma a)^2+\sigma^2\tau^2\right)^{1/2} \\ &\leq \sqrt{2}A\left(\sigma(a^2+\tau^2)^{N/2}+N(a^2+\tau^2)^{(N-1)/2}\right) \\ &\leq \sqrt{2}A(\sigma+N)(a^2+\tau^2)^{N/2},\end{aligned}$$

inequality (9.2.10) implies

(9.2.11) $$|\varphi'(\tau)| \leq CA(\sigma+N)(a^2+\tau^2)^{N/2}.$$

It is possible to extend this result to the multivariate case in the form: if $g \in B_{\sigma,m}$ satisfies

$$|g(x)| \leq A(1+|x|)^N, \qquad x \in \mathbb{R}^m$$

where $N \geq 0$ is an integer, then

(9.2.12) $$\left|\frac{\partial g(x)}{\partial x_k}\right| \leq CA(\sigma_k+N)(1+|x|)^N, \qquad x \in \mathbb{R}^m, \quad 1 \leq k \leq n.$$

Indeed, for each fixed $x_1,\ldots,x_{k-1},x_{k+1},\ldots,x_m$, the function $g(x)=\varphi(x_k)$ is an entire function in x_k of exponential type σ_k, satisfying the inequality

$$|\varphi(x_k)| \leq CA(1+x_1^2+\cdots+x_{k-1}^2+x_{k+1}^2+\cdots+x_m^2+x_k^2)^{N/2}, \quad x_k \in \mathbb{R}.$$

Then using (9.2.11), we get

$$\left|\frac{\partial g(x)}{\partial x_k}\right| = |\varphi'(x_k)| \leq AC(\sigma_k+N)(1+|x|^2)^{N/2} \leq AC(\sigma_k+N)(1+|x|)^N.$$

Thus (9.2.12) follows.

Next, we note that by the Mean Value Theorem,

(9.2.13) $$|g(x)-g((1-\varepsilon(N+2))x)| \leq C\varepsilon|x|\left(\sum_{k=1}^m\left(\frac{\partial g}{\partial x_k}\right)^2(\xi)\right)^{1/2},$$

where $\xi = \xi(x)$ satisfies the conditions

$$(9.2.14) \qquad (1 - \varepsilon(N+2))|x_k| \leq |\xi_k| \leq |x_k|, \qquad 1 \leq k \leq m.$$

Since g satisfies (9.1.9), we obtain from (9.2.12)–(9.2.14)

$$(9.2.15) \qquad \left|g(x) - g\big((1-\varepsilon(N+2))x\big)\right| \leq AC\varepsilon n^{m_2}(1+|x|)^{N+1}.$$

Thus (9.2.15) implies

$$(9.2.16) \qquad J_1 = \left\|g - g\big((1-\varepsilon(N+2))\cdot\big)\right\|_{L_{p,W}(\mathcal{I}_m)} \leq AC\varepsilon n^{m_3}.$$

Combining estimates (9.2.5), (9.2.9), and (9.2.16) with (9.2.4), we have

$$(9.2.17) \quad \begin{aligned} E_{\mu n,m}\big(g, L_{p,W}(\mathcal{I}_m)\big) &\leq \|g-g_1\|_{L_{p,W}(\mathcal{I}_m)} + E_{\mu n,m}\big(g_1, L_{p,W}(\mathcal{I}_m)\big) \\ &\leq AC\big(\varepsilon^2 n^{\gamma_3} + \varepsilon n^{m_3} + n^{\gamma_2}\varepsilon^{-m/2}\big)\exp\left(-C_1 n^{\min_{1\leq k \leq m} \delta_1(W_k)}\right). \end{aligned}$$

Finally, minimizing the right-hand side of (9.2.17) with respect to $\varepsilon \in (0, 1/(N+2))$, we obtain (9.1.10). ∎

9.3. Proofs of Theorems 9.1.1 and 9.1.4

Proof of Theorem 9.1.4. We use the induction in m. For $m = 1$ Theorem 9.1.4 follows from Theorem 2.3.2.

Assume that for any polynomial (9.1.11) from $\mathcal{P}_{\mu n, m}$, inequality (9.1.12) holds. Note that any polynomial $P \in \mathcal{P}_{\mu n, m+1}$ can be represented in the form

$$P(x_1, \ldots, x_m, x_{m+1}) = \sum_{0 \leq \ell \leq \mu_{m+1} n} P_\ell(x_1, \ldots, x_m) x_{m+1}^\ell,$$

where $P_\ell \in \mathcal{P}_{\mu n, m}$, $0 \leq \ell \leq \mu_{m+1} n$.

Then we have from Theorem 2.3.2

$$|P_\ell(x_1, \ldots, x_m)| \leq \frac{C\sqrt{\ell+1}\, b^\ell_{[\mu_{m+1} n]}(Q_{m+1})}{\ell!} \times \sup_{x_{m+1} \in [-a_{[\mu_{m+1} n]}, a_{[\mu_{m+1} n]}]} |P(x_1, \ldots, x_m, x_{m+1})| W_{m+1}(x_{m+1}).$$

Hence for $0 \leq \ell \leq \mu_{m+1} n$,

$$(9.3.1) \qquad \sup_{\mathcal{A}_{n,m}} |P_\ell| \prod_{i=1}^{m} W_i \leq \frac{C\sqrt{\ell+1}\, b^\ell_{[\mu_{m+1} n]}(Q_{m+1})}{\ell!} \sup_{\mathcal{A}_{n,m+1}} |P| \prod_{i=1}^{m+1} W_i.$$

Since every coefficient of P_ℓ has the form of $c_{k_1,\ldots,k_m,\ell}$, we deduce from assumption (9.1.12) and inequality (9.3.1) the following estimate

$$|c_{k_1,\ldots,k_m,k_{m+1}}| \leq C \prod_{i=1}^{m+1} \frac{\sqrt{k_i+1}\, b^{k_i}_{[\mu_i n]}(Q_i)}{k_i!} \|P\|_{L_{\infty,W}(\mathcal{A}_{n,m+1})}.$$

Therefore the theorem is established for $p = \infty$. Statement (b) of Theorem 9.1.4 can be proved similarly. ∎

Proof of Theorem 9.1.1. Let us set $\delta(W_k) := \delta_1(W_k)$, $1 \leq k \leq m$. Let $f \in L_\infty(\mathbb{R}^m)$. Then there exists $g \in B_{\sigma,m} \cap L_\infty(\mathbb{R}^m)$ such that

$$\|f - g\|_{L_\infty(\mathbb{R}^m)} = A_{\sigma,m}(f, L_\infty)$$

(see [101]). Hence

$$\limsup_{n \to \infty} E_{\mu n,m}\big(f(A_{\alpha(n)}\cdot), L_{\infty,W}(\mathcal{I}_m)\big)$$
$$= \limsup_{n \to \infty} E_{\mu n,m}\big(f(A_{\alpha(n)}\cdot) - g(A_{\alpha(n)}\cdot)), L_{\infty,W}(\mathcal{I}_m)\big)$$
$$+ \limsup_{n \to \infty} E_{\mu n,m}\big(g(A_{\alpha(n)}\cdot), L_{\infty,W}(\mathcal{I}_m)\big)$$

(9.3.2) $$\leq A_{\sigma,m}(f, L_\infty) + \limsup_{n \to \infty} E_{\mu n,m}\big(g(A_{\alpha(n)}\cdot), L_{\infty,W}(\mathcal{I}_m)\big).$$

Since by Theorem 9.1.3,

$$\lim_{n \to \infty} E_{\mu n,m}\big(g(A_{\alpha(n)}\cdot), L_{\infty,W}(\mathcal{I}_m)\big) = 0,$$

we obtain from (9.3.2)

(9.3.3) $$\limsup_{n \to \infty} E_{\mu n,m}\big(f(A_{\alpha(n)}\cdot), L_{\infty,W}(\mathcal{I}_m)\big) \leq A_{\sigma,m}(f, L_\infty).$$

Further let $P \in \mathcal{P}_{\mu n,m}$ be polynomial (9.1.11), satisfying

$$E_{\mu n,m}\big(f(A_{\alpha(n)}\cdot), L_{\infty,W}(\mathcal{A}_{n,m})\big) = \|f(A_{\alpha(n)}\cdot) - P\|_{L_{\infty,W}(\mathcal{A}_{n,m})}.$$

Hence we have

(9.3.4) $$\|P\|_{L_{\infty,W}(\mathcal{A}_{n,m})} \leq 2\|f\|_{L_\infty(\mathbb{R}^m)}.$$

Without loss of generality we may assume that

$$\lim_{n \to \infty} E_{\mu n,m}\big(f(A_{\alpha(n)}\cdot), L_{\infty,W}(\mathcal{A}_{n,m})\big)$$

exists. Next, by Theorem 9.1.4 and by (9.3.4),

(9.3.5) $$|c_{k_1,\ldots,k_m}| \leq C \prod_{i=1}^m \frac{\sqrt{k_i + 1}\, b_{[\mu_i n]}^{k_i}(W_i)}{k_i!} \|f\|_{L_\infty(\mathbb{R}^m)}.$$

Then setting

$$\Pi_n(x) = P(A_{\alpha(n)}^{-1} x) = \sum_{0 \leq k_i \leq \mu_i n,\, 1 \leq i \leq m} d_{k_1,\ldots,k_m} x_1^{k_1} \cdots x_m^{k_m},$$

where $A_{\alpha(n)}^{-1}$ is the inverse operator, we have from (9.3.5)

(9.3.6) $$|d_{k_1,\ldots,k_m}| \leq C \frac{\sqrt{k_i + 1}\, \sigma_i^{k_i}}{k_i!(1 - Cn^{-\delta(W_i)})^{k_i}}.$$

Using a multidimensional analogue of Proposition 5.1.2(a) (see [101]) to the sequence $\{\Pi_n\}_{n=1}^\infty$, we see that (9.3.6) implies the existence of a subsequence $\{n_s\}_{s=1}^\infty$ and a function g_0 such that

(9.3.7) $$\lim_{s \to \infty} \Pi_{n_s}(z) = g_0(z)$$

uniformly in each cube $K_m(M) = \{x \in \mathbb{R}^m : |x_i| \leq M,\, 1 \leq i \leq m\}$. Note that by (9.3.6), $g_0 \in B_{\sigma(1+\varepsilon),m}$ for any $\varepsilon > 0$.

Next, for every $\tau > 0$ there exists n_1 such that for $n > n_1$

$$\min_{y \in K_m(M)} W(A_{\alpha(n)}^{-1} y) = \min_{y \in K_m(M)} \exp\left(-\sum_{i=1}^{m} Q_i\left(\frac{\sigma_i(1 - Cn^{-\delta(W_i)})}{b_{[\mu_i n]}} y_i\right)\right)$$

(9.3.8)
$$\geq 1 - \tau.$$

Then for given $M > 0$ and $\tau \in (0, 1)$ and for n large enough we have from (9.3.8)

$$E_{\mu n, m}\big(f(A_{\alpha(n)}\cdot), L_{\infty, W}(\mathcal{A}_{n,m})\big) = \max_{x \in \mathcal{A}_{n,m}} |f(A_{\alpha(n)} x) - P(x)| W(x)$$

$$= \max_{y \in A_{\alpha(n)}(\mathcal{A}_{n,m})} |f(y) - \Pi_n(y)| W(A_{\alpha(n)}^{-1} y)$$

$$\geq \max_{y \in K_m(M)} |f(y) - \Pi_n(y)| W(A_{\alpha(n)}^{-1} y)$$

$$\geq (1 - \tau)\|f - \Pi_n\|_{L_\infty(K_m(M))}.$$

Hence by (9.3.7), we have

(9.3.9) $\quad \lim_{s \to \infty} E_{\mu n_s}\big(f(A_{\alpha(n_s)}\cdot), L_{\infty, W}(\mathcal{A}_{n_s, m})\big) \geq (1 - \tau)\|f - g_0\|_{L_\infty(K_m(M))}.$

Letting $M \to \infty$ and then $\tau \to 0$ in (9.3.9), we obtain

(9.3.10) $\quad \lim_{s \to \infty} E_{\mu n_s, m}\big(f(A_{\alpha(n_s)}\cdot), L_{\infty, W}(\mathcal{A}_{n_s, m}) \geq A_{\sigma(1+\varepsilon), m}(f, L_\infty).$

Further taking account of the relation

$$\lim_{\varepsilon \to 0+} A_{\sigma(1+\varepsilon), m}(f, L_\infty) = A_{\sigma, m}(f, L_\infty)$$

(see [47; Lemma 11.11]), we obtain by (9.3.10)

(9.3.11) $\quad \liminf_{n \to \infty} E_{\mu n, m}\big(f(A_{\alpha(n)}\cdot), L_{\infty, W}(\mathcal{A}_{n,m})\big) \geq A_{\sigma, m}(f, L_\infty).$

Thus (9.3.3) and (9.3.11) yield (9.1.4) for $f \in L_\infty(\mathbb{R}^m)$. The proof of relation (9.1.4) for a function, satisfying (9.1.5) and (9.1.6), is similar to that of Theorem 2.1.1(b) if we use Theorems 9.1.1(a) and 9.1.3 instead of Theorems 2.1.1(a) and 2.2.1. ∎

The proof of Theorem 9.1.2 is similar to those of Theorems 2.1.2 and 9.1.1, so it is omitted.

9.4. Approximation of λ-Homogeneous Functions

Here, we consider the case of "cubic" spectra and equal weights W_i, i.e.,

(9.4.1) $\quad \sigma = (\sigma_1, \ldots, \sigma_1), \ \mu = (\mu_1, \ldots, \mu_1), \ W(x) = \exp\left(-\sum_{k=1}^{m} Q(x_k)\right).$

Let f_λ be continuous λ-homogeneous function on $\mathbb{R}^m$ with $\lambda > 0$. Since

$$|f_\lambda(x)| \leq A|x|^\lambda, \qquad x \in \mathbb{R}^m,$$

the function f_λ satisfies condition (9.1.5) for $N = \lambda$. Thus the following result is an immediate consequence of Theorems 9.1.1(b) and 9.1.2(b):

114 9. MULTIDIMENSIONAL LIMIT THEOREMS OF POLYNOMIAL APPROXIMATION

THEOREM 9.4.1. *Let σ, μ, and W satisfy* (9.4.1) *and let* $\exp(-Q(\tau)) \in \mathcal{F}(C^2)$. *If f_λ, $\lambda > 0$, satisfies the condition*

(9.4.2) $$A_{\sigma_0,m}(f_\lambda, L_p) < \infty$$

for some $\sigma_0 = (\sigma_{0,1}, \ldots, \sigma_{0,1})$ with $\sigma_{0,1} > 0$ and for $p \in [1, \infty]$, then for all $\sigma_1 > 0$,

(9.4.3) $$\lim_{n \to \infty} b_{[\mu_1 n]}^{\lambda + m/p} E_{\mu n, m}(f_\lambda, L_{p,W}(\mathcal{I}_m)) = \sigma_1^{\lambda + m/p} A_{\sigma,m}(f_\lambda, L_p) < \infty.$$

This theorem shows that asymptotic (9.4.3) holds if condition (9.4.2) is satisfied. In Section 8.1, we proved that for $m = 1$ (9.4.2) is valid for any $\lambda > 0$ and $p \in (0, \infty]$. This is not true for $m > 1$. For example, if $f_\lambda(x) = |x_1|^{\lambda_1} \cdots |x_m|^{\lambda_m}$, where $0 \le \lambda_1 \le \cdots \le \lambda_m < \lambda$, $\sum_{i=1}^{m} \lambda_i = \lambda$ and for some k_0, $1 \le k_0 \le m$, $\lambda_{k_0} \ne 0, 2, \ldots$, then $A_{\sigma,m}(f_\lambda, L_\infty) = \infty$ (see [40]).

A criterion for

(9.4.4) $$A_{\sigma,m}(f_\lambda, L_\infty) < \infty, \qquad \lambda > 0,$$

given in [40], states that (9.4.4) holds if and only if f_λ belongs to the Nikolskii class $H_\infty^\lambda(\Omega)$ in some neighborhood Ω of the origin. It is possible to extend this criterion to $p \in [1, \infty]$ if $H_\infty^\lambda(\Omega)$ is replaced with $H_p^\lambda(\Omega)$ (see [101] for the definitions).

In the following proposition, we give simple sufficient conditions on f for

(9.4.5) $$A_{\sigma,m}(f, L_p) < \infty.$$

PROPOSITION 9.4.2. *Let $f \in L_{1,\mathrm{loc}}(\mathbb{R}^m)$ and let the Fourier transform $\hat{f}$ of the tempered distribution f is a k-differentiable function outside of the parallelepiped $\prod_\varepsilon := \{x \in \mathbb{R}^m \colon |x_i| \le \varepsilon_i, 1 \le i \le m\}$, $\sum_{i=1}^{m} \varepsilon_i > 0$, with*

$$D^\ell \hat{f} := \frac{\partial^{|\ell|} \hat{f}}{\partial^{\ell_1} x_1 \cdots \partial^{\ell_m} x_m} \in L_1(\mathbb{R}^m \setminus \prod_\varepsilon)$$

and $\lim_{|x| \to \infty} D^\ell \hat{f}(x) = 0$, where $\ell = (\ell_1, \ldots, \ell_m)$, $0 \le |\ell| := \ell_1 + \cdots + \ell_m \le k$. Then for $p > m/k$ and $\sigma_i \ge \varepsilon_i$, $1 \le i \le m$, inequality (9.4.5) holds with $\sigma = (\sigma_1, \ldots, \sigma_m)$.

PROOF. By Whitney's theorem [85] there is an extension $F_{k,\varepsilon}$ of $\hat{f}$ from $\mathbb{R}^m \setminus \prod_\varepsilon$ to $\mathbb{R}^m$ such that for all multiindices ℓ with $0 \le |\ell| \le k$, $D^\ell F_{k,\varepsilon} \in L_1(\mathbb{R}^m)$. Moreover, we have for the inverse Fourier transform

(9.4.6) $$|\check{f}_{k,\varepsilon}(x)| \le C(1 + |x|)^{-k}, \qquad k \ge 1.$$

Next, the function $H(x) := (\hat{f} - F_{k,\varepsilon})(x)$ is a tempered distribution with the support in $\prod_\varepsilon$, so by the generalized Paley-Wiener theorem [40], the function $g_\varepsilon = \check{H} = f - \check{F}_{k,\varepsilon}$ belongs to $B_{\sigma,m}$. Since by (9.4.6), $f - g_\varepsilon \in L_p(\mathbb{R}^m)$ for $p > m/k$, (9.4.5) follows. ∎

For $m = 1$ the similar conditions are considered in Lemma 8.1.5. Next, we apply Proposition 9.4.2 to the λ-homogeneous function $f_{\lambda,m}(x) := |x|^\lambda = \left(\sum_{i=1}^{m} x_i^2\right)^{\lambda/2}$, $\lambda > 0$. Since the Fourier transform

$$\hat{f}_{\lambda,m}(x) = C(\lambda, m)|x|^{-\lambda - m}$$

9.4. APPROXIMATION OF λ-HOMOGENEOUS FUNCTIONS

(see [20]) satisfies all conditions of Proposition 9.4.2 for $k = 1, 2, \ldots$, and any ε with $\prod_{i=1}^{m} \varepsilon_i > 0$, we have that

$$A_{\sigma,m}(f_{\lambda,m}, L_p) < \infty, \qquad 0 < p \leq \infty, \quad \lambda > 0.$$

Thus the following corollary immediately follows from Theorem 9.4.1:

COROLLARY 9.4.3. *Let σ, μ, and W satisfy* (9.4.1) *and let* $\exp(-Q(\tau)) \in \mathcal{F}(C^2)$. *Then for $\sigma_1 > 0$, $\mu_1 > 0$ and $p \in [1, \infty]$,*

$$\lim_{n \to \infty} b_{[\mu,n]}^{\lambda+m/p} E_{\mu n,m}(f_{\lambda,m}, L_{p,W}(\mathcal{I}_m)) = \sigma_1^{\lambda+m/p} A_{\sigma,m}(f_{\lambda,m}, L_p) < \infty.$$

CHAPTER 10

Examples

In this chapter, we discuss the limit theorems for the "canonical" weights $W_\alpha(x) := \exp(-|x|^\alpha)$, $\alpha > 0$. In particular, for $\alpha > 1$ the constant β in Theorems 2.1.3 and 2.2.2 is computed. For $\alpha = 1$ we obtain different versions of the limit theorems since $W_1 \notin \mathcal{F}(C^2)$. In the case $0 < \alpha < 1$, a description of the class of functions f for which $\lim_{n \to \infty} E_n(f, L_{\infty, W_\alpha}(\mathbb{R})) = 0$ is given. We also show that the error of non-weighted polynomial approximation on $[-1, 1]$ of a function is the limit as $\alpha \to \infty$ of the corresponding weighted errors associated with the weight W_α. Examples of Erdös weights on bounded and unbounded intervals are discussed as well.

10.1. $W(x) = \exp(-|x|^\alpha)$, $\alpha > 1$

The weight

$$W_\alpha(x) := \exp\bigl(-Q_\alpha(x)\bigr),$$

where

$$Q_\alpha(x) := |x|^\alpha, \qquad \alpha > 0, \quad x \in I = \mathbb{R},$$

is apparently the most studied exponential weight in weighted approximation and orthogonal polynomials. In particular, W_2 is the classical Hermite weight.

It is easy to see that for $\alpha > 1$, $W_\alpha \in \mathcal{F}(C^2)$ with

(10.1.1) $$T(x) = \Lambda = \alpha, \qquad x \in \mathbb{R},$$

and for $\alpha > 0$, $W_\alpha \in \mathcal{F}(C^3)$ with $C_1 = C_2 = \alpha$, $C_3 = (\alpha - 1)|\alpha - 2|$. Also, $W_\alpha \in \mathcal{F}_M$, $\alpha > 0$.

The constants a_n, b_n and the functions $H(ir)$, $h(r)$, $\phi(r)$, and $\phi_1(r)$ associated with W_α, $\alpha > 0$, are defined at (1.4.12), (1.2.2), (3.2.1), (3.3.1), and (3.3.16) for $W = W_\alpha$, respectively. They can be expressed in terms of the gamma function $\Gamma(z)$ or the hypergeometric function $F(a, b, c, z)$ as follows:

PROPOSITION 10.1.1. *Let us set*

(10.1.2) $$B_\alpha := \frac{\alpha \Gamma((\alpha+1)/2)}{\sqrt{\pi}\, \Gamma(\alpha/2 + 1)} = \frac{\Gamma(\alpha)}{2^{\alpha-2} \Gamma^2(\alpha/2)}, \qquad \alpha > 0.$$

Then for $Q = Q_\alpha$ the following formulae hold:

(10.1.3)
$$a_n = a_n(Q_\alpha) = B_\alpha^{-1/\alpha} n^{1/\alpha}, \quad \alpha > 0,$$
$$b_n = b_n(Q_\alpha) = \frac{\alpha B_\alpha}{\alpha - 1} a_n^{\alpha - 1} = \frac{\alpha n}{(\alpha - 1)a_n}$$

(10.1.4)
$$= \frac{\alpha B_\alpha^{1/\alpha} n^{\frac{\alpha-1}{\alpha}}}{\alpha - 1}, \quad \alpha > 1,$$

$$H(ir) = \frac{2n}{\alpha} \left(F\left(-\frac{\alpha}{2}, 1, \frac{\alpha}{2} + 1, r^2\right) - \frac{1}{2} \right),$$

(10.1.5)
$$0 \leq r < 1, \quad \alpha > 0,$$

$$h(r) = \frac{2r\alpha}{a_n(1 - r^2)} H(ir) = \frac{4nr}{a_n(1 - r^2)}$$

(10.1.6)
$$\times \left(F\left(-\frac{\alpha}{2}, 1, \frac{\alpha}{2} + 1, r^2\right) - \frac{1}{2} \right), \quad 0 \leq r < 1, \quad \alpha > 0,$$

$$\phi(r) = \frac{a_n h(r)(1 - r^2)}{2r} - H(ir) + n \log r$$
$$= (\alpha - 1) H(ir) + n \log r$$
$$= n \left(\frac{2(\alpha - 1)}{\alpha} \left(F\left(-\frac{\alpha}{2}, 1, \frac{\alpha}{2} + 1, r^2\right) - \frac{1}{2} \right) + \log r \right),$$

(10.1.7)
$$0 < r < 1, \quad \alpha \geq 1,$$

$$\phi_1(r) = \frac{a_n h(r)(1 + r^2)^2}{2r(1 - r^2)} - H(ir) + n \log r$$
$$= \frac{\alpha(1 + r^2)^2 - (1 - r^2)^2}{(1 - r^2)^2} H(ir) + n \log r$$
$$= n \left(\frac{2(\alpha(1 + r^2)^2 - (1 - r^2)^2)}{\alpha(1 - r^2)^2} \left(F\left(-\frac{\alpha}{2}, 1, \frac{\alpha}{2} + 1, r^2\right) - \frac{1}{2} \right) + \log r \right),$$

(10.1.8)
$$0 < r < 1, \quad \alpha \geq 1.$$

PROOF. We first note that the identity in (10.1.2) follows from the double-argument formula for Γ [32, Eq. (1.2.(15))]. Next, relation (10.1.3) follows from (1.4.12) and (10.1.2) (cf. [93]). Formulae (10.1.4) for b_n can be deduced from (1.2.2), (10.1.2), and (10.1.3). Relation (10.1.5) for $H(ir)$ was established in [93] in a slightly different form.

To prove (10.1.6), we recall that $W_\alpha \in \mathcal{F}_M$, $\alpha > 0$, and use (3.1.3) and (3.2.2) for $0 \leq r < 1$.

(10.1.9) $H(ir) = \dfrac{a_n^\alpha(1 - r^2)}{2\pi} \displaystyle\int_{-\pi}^{\pi} \dfrac{|\cos t|^\alpha}{1 + r^2 + 2r \sin t} \, dt$

$= \dfrac{a_n^\alpha(1 - r^2)}{\pi} \displaystyle\int_0^\pi \dfrac{|\sin t|^\alpha}{1 + r^2 + 2r \cos t} \, dt = \dfrac{a_n(1 - r^2)}{2r\alpha} h(r).$

Thus (10.1.5) and (10.1.9) yield (10.1.6). Finally, (10.1.7) follows from (3.3.1), (10.1.5), and (10.1.6), while (10.1.8) is a consequence of (3.3.16), (10.1.5), and (10.1.6). □

We remark that the restriction $\alpha > 1$ in (10.1.4) is essential since $b_n = \infty$ if $0 < \alpha \leq 1$, while formulae (10.1.7) and (10.1.8) are also valid for $\alpha \in (0,1)$.

To evaluate the constant β in Theorems 2.1.3 and 2.2.2 for $W = W_\alpha$, $\alpha > 1$, we define $r_0 = r_0(\alpha)$ to be the only solution to the equation

$$(10.1.10) \qquad F\left(-\frac{\alpha}{2}, 1, \frac{\alpha}{2}+1, r^2\right) = \frac{1}{2} + \frac{(1-r^2)^2 \log(1/r)}{2((1+r^2)^2 - (1-r^2)/\alpha)}$$

in the interval $(0,1)$. Next, let us set

$$(10.1.11) \qquad \beta(\alpha) := \left(\frac{2(\alpha-1)r_0(1+r_0^2)}{\alpha(1+r_0^2) - (1-r_0^2)^2} \log \frac{1}{r_0}\right)^{-1}.$$

THEOREM 10.1.2. *Let* $W = W_\alpha$, $\alpha > 1$. *Then the following statements hold:*

(a) *There exist* $\delta(W) \in \left(0, \frac{2(\alpha-1)}{3(\alpha+1)}\right)$ *and* $C = C(W)$ *such that for any* $f \in L_{\infty,\mathrm{loc}}(\mathbb{R})$, *satisfying (2.1.5) for some* $\sigma_0 > 0$, *and for every* $\sigma > \beta(\alpha)\sigma_0$, *(2.1.2) is valid.*

(b) *For any* $f \in L_{p,\mathrm{loc}}(\mathbb{R})$, $0 < p < \infty$, *satisfying (2.1.8) for some* $\sigma_0 > 0$, *and for every* $\sigma > \beta(\alpha)\sigma_0$, *(2.1.7) is valid.*

(c) *For any* $\varepsilon > 0$ *small enough, every* $\sigma \in (0, b_n/(\beta(\alpha)(1+\varepsilon))$ *and for all* $g \in B_\sigma$, *inequality (2.2.4) is valid.*

PROOF. We first note that (10.1.10) is equivalent to the equation

$$(10.1.12) \qquad \phi_1(r) = 0.$$

Then by Proposition 3.3.6(d) there exists the unique solution $r_0 = r_0(\alpha)$ to equation (10.1.12). Next using (10.1.4), (10.1.6), (10.1.10), and (10.1.11), we have

$$\sigma_0(\alpha) := \frac{h(r_0)(1+r_0^2)}{1-r_0^2} = \frac{4nr_0(1+r_0^2)}{a_n(1-r_0^2)^2}$$
$$\times \left(F\left(-\frac{1}{2}, 1, \frac{\alpha}{2}+1, r_0^2\right) - \frac{1}{2}\right)$$
$$= \frac{2nr_0(1+r_0^2)}{a_n((1+r_0^2)^2 - (1-r_0^2)/\alpha)} \log \frac{1}{r_0}$$
$$(10.1.13) \qquad = \frac{2(\alpha-1)r_0(1+r_0^2)b_n}{\alpha(1+r_0^2)^2 - (1-r_0^2)^2} \log \frac{1}{r_0} = b_n/\beta(\alpha).$$

Let now $g \in B_\sigma$, where $\sigma \in (0, \sigma_0(\alpha)/(1+\varepsilon)]$ and let $\varepsilon > 0$ be small enough. Then by (6.4.10),

$$E_n(g, L_p[-a_n, a_n]) \leq C(g, \varepsilon) n^\gamma \exp(-C_1 \varepsilon n).$$

This proves statement (c) of the theorem. Finally, using Theorem 10.1.2(c) instead of Theorem 2.2.2(a) in the proof of Theorem 2.1.3, we arrive at statements (a) and (b). ∎

For the weight W_α, $\alpha > 1$, it is possible to specify the constant b_n in the limit relations discussed in Chapter 8. In particular, the following result is a special case of Theorem 8.1.1(b):

COROLLARY 10.1.3. *For $f_\lambda(x) := |x|^\lambda$, $p \in (0, \infty]$, $\lambda > \max(-1, -1/p)$, and $\alpha > 1$, we have*

$$\lim_{n \to \infty} n^{\frac{(\alpha-1)(\lambda+1/p)}{\alpha}} E_n(f_\lambda, L_{p,W_\alpha}(\mathbb{R}))$$
$$= \lim_{n \to \infty} n^{\frac{(\alpha-1)(\lambda+1/p)}{\alpha}} E_n(f_\lambda, L_{p,W_\alpha}[-B_\alpha^{-1/\alpha} n^{1/\alpha}, B_\alpha^{-1/\alpha} n^{1/\alpha}])$$
$$= [(\alpha-1)/(\alpha B_\alpha^{1/\alpha})]^{\lambda+1/p} B_{\lambda,p}.$$

Remarks

(a) Relations (3.2.6) and (10.1.3) show that
$$\inf a_n(Q) b_n(Q) = n, \qquad \sup a_n(Q) b_n(Q) = \infty,$$
where the lower and upper bounds are taken over all $W(x) = \exp(-Q(x)) \in \mathcal{F}$.

(b) We conjecture that the solution $r_0(\alpha)$ to (10.1.10) is decreasing in $\alpha \in (1, \infty)$. If it is true, then $r_0(\alpha) \in (r_0(\infty), r_0(1))$, where $r_0(1) = \sqrt{2} - 1 = 0.41421\ldots$ is the solution to the equation
$$F\left(-\frac{1}{2}, 1, \frac{3}{2}, r^2\right) = \frac{1}{2} + \frac{(1-r^2)^2}{8r^2} \log \frac{1}{r},$$
and $r_0(\infty) = 0.30129\ldots$ is the solution to the equation
$$(10.1.14) \qquad \frac{1+r^2}{1-r^2} = \log \frac{1}{r}.$$
We recall that $r_0(\alpha) \in [0.24, 0.60]$, by Proposition 3.3.8. To find $r_0(\infty)$, it suffices to note that
$$\lim_{\alpha \to \infty} F\left(-\frac{\alpha}{2}, 1, \frac{\alpha}{2}+1, r^2\right) = (1+r^2)^{-1}$$
uniformly in $r \in [0.24, 0.60]$.

(c) It easily follows from (10.1.11) and (10.1.14) that
$$\beta(\infty) := \lim_{\alpha \to \infty} \beta(\alpha) = \frac{1 - r_0^2(\infty)}{2 r_0(\infty)} = 1.50887\ldots.$$
Moreover, by the substitution $\tau = (1-r^2)/(2r)$, we reduce (10.1.14) to Bernstein's equation (1.1.8). Thus $\beta(\infty)$ coincides with Bernstein's constant (1.1.7) for nonweighted polynomial approximation on $[-1, 1]$ (cf. Theorem 10.4.1).

Bernstein [4, 16] showed that there exists $g \in B_\sigma$ such that
$$\lim_{n \to \infty} E_n\left(g\left(\frac{n(1-\lambda_n)}{\sigma} \cdot\right), L_\infty[-1, 1]\right) = \infty,$$
for all $\sigma \in (0, \beta(\infty))$ and $\lim_{n \to \infty} \lambda_n = 0$. We conjecture that $\beta(\alpha)$ is the best possible constant in Theorem 10.1.2(a), i.e., there exists $g \in B_\sigma$ such that
$$\lim_{n \to \infty} E_n\left(g\left(\frac{b_n(1-Cn^{-\delta})}{\sigma}\right), L_{\infty, W_\alpha}[-a_n, a_n]\right) = \infty$$
for all $\sigma \in (0, \beta(\alpha))$.

(d) It is possible to show that for $\alpha > 3$,
$$\phi(r) = -C(\alpha) n (1-r)^3 (1 + o(1)), \qquad r \to 1-,$$
where ϕ is given in (10.1.7). Thus the exponent 3 in (3.3.2) cannot be replaced by a smaller number for all weights from $\mathcal{F}_M$.

10.2. $W(x) = \exp(-|x|)$

The weight $W_1(x) := \exp(-Q_1(x))$, where $Q_1(x) := |x|$, does not belong to $\mathcal{F}(C^2)$ since $T(x) = 1$ and condition (1.4.5) fails. Moreover, $b_n(Q_1) = \infty$, so limit relations (2.1.1), (2.1.2), (2.1.6), and (2.1.7) are not valid for $W = W_1$. Nevertheless, using the fact that $W_1 \in \mathcal{F}_M$, we obtain some limit estimates like (7.1.1) and (7.1.2) with b_n replaced by $C \log n$, where $C = 2/(3\pi)$ for the upper estimate and $C = 2/\pi$ for the lower one. As a corollary, we establish limit estimates for $E_n(f_\lambda, L_{\infty,W_1}(\mathbb{R}))$.

THEOREM 10.2.1. *Let a continuous function f satisfy the conditions (2.1.4) and (2.1.5) for some $\sigma_0 > 0$. Then the following statements hold:*

(a) *For $\sigma \geq \sigma_0$ and any $\varepsilon \in (0,1)$,*

(10.2.1) $$\limsup_{n \to \infty} E_n\bigl(f(\gamma_n^* \cdot), L_{\infty,W_1}(\mathbb{R})\bigr) \leq A_\sigma(f, L_\infty),$$

where
$$\gamma_n^* := \frac{2(1-\varepsilon)}{3\pi\sigma} \log n.$$

(b) *For $\sigma > 3\sigma_0$,*

(10.2.2) $$\liminf_{n \to \infty} E_n\bigl(f(\gamma_n^{**} \cdot), L_{\infty,W_1}[-a_n, a_n]\bigr) \geq A_\sigma(f, L_\infty),$$

where
$$\gamma_n^{**} := \frac{2}{\pi\sigma} \log n.$$

The chief ingredients of the proof are the following analogues of Theorems 2.2.1 and 2.3.1:

LEMMA 10.2.2. *For any $\varepsilon \in (0,1)$, $\sigma \in \bigl(0, \frac{2(1-\varepsilon)}{3\pi} \log n\bigr)$ and for every $g \in B_\sigma$, satisfying (2.2.2), we have*

(10.2.3) $$E_n\bigl(g, L_{\infty,W_1}(\mathbb{R})\bigr) \leq C A n^\gamma \exp(-C_1 n^\varepsilon).$$

LEMMA 10.2.3. *For any polynomial $P(x) = \sum_{k=0}^{n} c_k x^k$, the following inequalities hold:*

(10.2.4) $\quad |P(z)| \leq e^{\mu_n |z|} \|P\|_{L_{\infty,W_1}[-a_n,a_n]}, \quad |z| = |x + iy| \geq 1/\mu_n,$

(10.2.5) $\quad |c_k| \leq \dfrac{3\sqrt{k+1}}{k!} \mu_n^k \|P\|_{L_{\infty,W_1}[-a_n,a_n]}, \quad 0 \leq k \leq n, \quad n \geq 2,$

where

(10.2.6) $$\mu_n := (2/\pi)(\log n + \log(\log n) + o(\log(\log n))), \quad n \to \infty.$$

To prove these lemmas, we first need the following version of Proposition 10.1.1 for $\alpha = 1$:

LEMMA 10.2.4. *The following formulae hold:*

(10.2.7) $$a_n = a_n(Q_1) = \pi n/2,$$

$$H(w) = \frac{4a_n}{\pi}\left(F\left(-\frac{1}{2},1,\frac{3}{2},-w^2\right) - \frac{1}{2}\right)$$

(10.2.8) $$= \frac{a_n(w^2+1)}{\pi i w}\log\left(\frac{1+iw}{1-iw}\right), \quad w \in \mathbb{C}, \ |w| < 1,$$

(10.2.9) $$H(ir) = \frac{a_n(1-r^2)}{\pi r}\log\left(\frac{1+r}{1-r}\right), \quad r \in [0,1),$$

(10.2.10) $$h(r) = \frac{2}{\pi}\log\left(\frac{1+r}{1-r}\right), \quad r \in [0,1),$$

(10.2.11) $$\phi(r) = n\log r, \quad r \in (0,1).$$

PROOF. We first note that (10.2.7) follows from (10.1.3), (10.2.11) is a consequence of (10.1.7), (10.2.8) implies (10.2.9), and (10.2.10) follows from (10.1.6) and (10.2.9).

It remains to prove (10.2.8). Indeed, using the Mhaskar-Saff representation for $H(W_1, w)$ [93] and taking account of the well-known relations for the hypergeometric functions [32, Eqs. (2.8.(28)), (2.8.(14)), (2.8.(4))], we obtain for $0 < |w| < 1$

$$(\pi/a_n)H(w) = 4\left(F\left(-\frac{1}{2},1,\frac{3}{2},-w^2\right) - \frac{1}{2}\right)$$
$$= 2\left((1+w^2)F\left(\frac{1}{2},1,\frac{3}{2},-w^2\right) + (1+w^2)F\left(\frac{3}{2},1,\frac{3}{2},-w^2\right) - 1\right)$$
$$= \frac{2(w^2+1)}{w}\arctan w = \frac{w^2+1}{iw}\log\left(\frac{1+iw}{1-iw}\right).$$

Hence (10.2.8) follows. ∎

Proof of Lemma 10.2.3. Since estimate (3.4.1) with b_n replaced by any constant and $W = W_1$ does not hold in a neighborhood of $\pm i$, we first obtain an upper bound for Re $H(w)$, where $w \in \mathbb{C}$ satisfies the conditions

$$|w| \leq 1, \quad (1/2)|w + w^{-1}| \geq \gamma.$$

Here γ is a positive number.

It immediately follows from (10.2.8) that

(10.2.12) $$\text{Re } H(w) \leq |H(w)| \leq (a_n/\pi)|w + w^{-1}|\left(\left|\log\left|\frac{1+iw}{1-iw}\right|\right| + 2\pi\right).$$

Next, for $0 \leq |w| < 1/2$,

(10.2.13) $$\left|\log\left|\frac{1+iw}{1-iw}\right|\right| < \log 4,$$

and for $1/2 \leq |w| \leq 1$,

(10.2.14) $$\left|\log\left|\frac{1+iw}{1-iw}\right|\right| \leq \left|\log\frac{(1+|w|)^2}{|w+w^{-1}|}\right| \leq \left|\log\left(\frac{2}{|w+w^{-1}|}\right)\right| + \log 4.$$

Further for $0 \leq |w| \leq 1$,

(10.2.15) $$\log\frac{1}{|w|} \leq \frac{1-|w|^2}{2|w|} \leq \frac{1}{2}|w + w^{-1}|.$$

Collecting inequalities (10.2.12)–(10.2.15) and taking account of (10.2.7), we have for $0 \leq |w| \leq 1$ and $(1/2)|w + w^{-1}| \geq \gamma$, where $\gamma \in (0,1)$,

$$\text{(10.2.16)} \qquad \operatorname{Re} H(w) + n \log \frac{1}{|w|} \leq \frac{a_n}{\pi}\left(\log \frac{1}{\gamma} + C\right)|w + w^{-1}|.$$

Here $C = \log 4 + 2\pi + 1$. Let us choose $\gamma = \gamma_n$, where $\gamma_n \in (0, 1/n)$, $n \geq 2$, is the unique solution to the equation

$$\text{(10.2.17)} \qquad \log \frac{1}{\gamma} + C = \frac{1}{n\gamma}.$$

Then
$$\mu_n := \frac{2}{\pi n \gamma_n} = \frac{2}{\pi}\big(\log n + \log(\log n) + o(\log(\log n))\big), \qquad n \to \infty$$

(cf. [35, p. 50]).

Furthermore, making the substitution $z = (a_n/2)(w + w^{-1})$, we have from (10.2.16) and (10.2.17) that for $|z| \geq a_n \gamma_n = 1/\mu_n$,

$$\operatorname{Re} H\left(\frac{1}{\psi(z/a_n)}\right) + n \log\left(\frac{1}{|\psi(z/a_n)|}\right)$$
$$\text{(10.2.18)} \qquad \leq \frac{2}{\pi}\left(\log \frac{1}{\gamma_n} + C\right)|z| = \mu_n |z|.$$

Using now inequality (4.3.5) for $W_1 \in \mathcal{F}_M$, we obtain from (10.2.18) that the estimate
$$|P(z)| \leq \exp(\mu_n |z|) \|P\|_{L_\infty, W_1[-a_n, a_n]}$$
holds for all $z \in \mathbb{C}$ with $|z| \geq 1/\mu_n$, $n = 2, 3, \ldots$. Thus (10.2.4) follows.

The proof of (10.2.5) is similar to that of Theorem 2.3.2. Note first that (10.2.5) holds for $k = 0$. Next for any $r \geq 1/\mu_n$ and $k \geq 1$, we have from (10.2.4)

$$|c_k| = \frac{1}{2\pi}\left|\int_{|z|=r} \frac{P(z)}{z^{k+1}}\, dz\right| \leq r^{-k} \max_{|z|=r} |P(z)|$$
$$\text{(10.2.19)} \qquad \leq \exp(\mu_n r - k \log r)\|P\|_{L_\infty, W_1[-a_n, a_n]}, \quad 1 \leq k \leq n,$$

where the asymptotic for μ_n is given in (10.2.6). Choosing $r = k/\mu_n \geq 1/\mu_n$, we have by (10.2.19) and by Stirling's formula,

$$|c_k| \leq \frac{\mu_n^k}{(k/e)^k}\|P\|_{L_\infty, W_1[-a_n, a_n]} \leq \frac{3\sqrt{k+1}\,\mu_n^k}{k!}\|P\|_{L_\infty, W_1[-a_n, a_n]}.$$

This yields (10.2.5). ∎

Proof of Lemma 10.2.2. Similarly to the proof of Theorem 2.2.1, we consider the approximation of $g \in B_\sigma$ by the Lagrange interpolation polynomials $P_{n-1} \in \mathcal{P}_{n-1}$ to g at the zeros $\{x_{k,n}^{(2)}\}_{k=1}^n$ of the orthogonal polynomials $T_{n,2}$ associated with W_1^2. Then

$$\|g - P_{n-1}\|_{L_\infty, W_1(\mathbb{R})} \leq \|g - P_{n-1}\|_{L_\infty, W_1(I_n)} + \|g - P_{n-1}\|_{L_\infty, W_1(\mathbb{R}\setminus I_n)}$$
$$\text{(10.2.20)} \qquad = J_1 + J_2,$$

where $I_n = [-a_n(1+\delta_n), a_n(1+\delta_n)]$ and $\delta_n := Cn^{-2(1-\varepsilon)/3}$, $n = 1, 2, \ldots$. Note that $\{\delta_n\}_{n=1}^\infty$ satisfies (6.0.16) if $n > C_2$.

We first estimate J_1. Using Proposition 6.1.3 for $\varepsilon_1 = \varepsilon/3$, we have that for any R, satisfying
$$R \geq \max\bigl(1 + 5\sqrt{\delta_n}, 1 + C_1 n^{-(1-\varepsilon)/3}\bigr), \qquad n > C_2,$$
the inequality
$$(10.2.21) \qquad J_1 \leq C n^\gamma R^{-n} e^{-H(i/R)} \max_{|w|=R} \left| g\left(\frac{a_n}{2}(w + 1/w)\right) \right|$$
holds. Since g satisfies (2.2.2), we obtain by (10.2.21) and (5.1.3),
$$(10.2.22) \qquad J_1 \leq AC n^\gamma r^{-N-1} \exp(S_\sigma(r)),$$
where $r \in (0,1)$ is a number satisfying
$$(10.2.23) \qquad r \leq \min\left(1 - \frac{5\sqrt{\delta_n}}{1 + 5\sqrt{\delta_n}}, 1 - C_3 n^{-(1-\varepsilon)/3}\right) = 1 - C_4 n^{-(1-\varepsilon)/3},$$
and $S_\sigma(r)$ is defined by (6.1.17).

Next by (3.2.1), S_σ satisfies the equation
$$\frac{dS_\sigma(r)}{dr} = \frac{a_n(1+r^2)}{2r^2}(h(r) - \sigma),$$
where h is given in (10.2.10) and by Proposition 3.2.1(c), it is a continuous, positive and increasing function in $r \in (0,1)$. Hence
$$(10.2.24) \qquad \begin{aligned} \min_{r\in(0,1)} S_\sigma(r) &\leq \min_{r\in(0,1)} S_{\frac{2(1-\varepsilon)}{3\pi}\log n}(r) \\ &= S_{\frac{2(1-\varepsilon)}{3\pi}\log n}(r_n) = \phi(r_n), \end{aligned}$$
where ϕ is defined in (10.2.11) and r_n is the solution to the equation
$$h(r_n) = \frac{2}{\pi}\log\left(\frac{1+r_n}{1-r_n}\right) = \frac{2(1-\varepsilon)}{3\pi}\log n.$$
Then
$$r_n = \bigl(1 - n^{-(1-\varepsilon)/3}\bigr) / \bigl(1 + n^{-(1-\varepsilon)/3}\bigr) \leq 1 - n^{-(1-\varepsilon)/3},$$
and
$$(10.2.25) \qquad \exp(\phi(r_n)) = \exp(n\log r_n) \leq \bigl(1 - n^{-(1-\varepsilon)/3}\bigr)^n \leq C\exp(-n^{2/3}).$$

Since r_n satisfies (10.2.23) for $C_4 = 1$, we have from (10.2.22), (10.2.24), and (10.2.25)
$$(10.2.26) \qquad \begin{aligned} J_1 &\leq CAn^\gamma \min_{r\in(0, 1-n^{-(1-\varepsilon)/3})} r^{-N-1} \exp(S_\sigma(r)) \\ &\leq ACn^\gamma \exp(-n^{-2/3}). \end{aligned}$$

Further by Proposition 6.2.2 and by relations (2.2.2), (10.2.26), we get
$$(10.2.27) \qquad \begin{aligned} J_2 &\leq Cn^{m_1}\bigl(\exp(-C_2\delta_n^{3/2}) + \exp(-C_1 n)\bigr)(J_1 + A) \\ &\leq Cn^{m_1}\exp(-C_5 n^\varepsilon). \end{aligned}$$

Thus (10.2.20), (10.2.26), and (10.2.27) yield (10.2.3). ∎

Proof of Theorem 10.2.1. The proof is similar to that of Theorem 2.1.1. Let first $f \in L_\infty(\mathbb{R})$. Then by Proposition 5.2.1 there exists $g \in B_\sigma \cap L_\infty(\mathbb{R})$ such that $\|f - g\|_{L_\infty(\mathbb{R})} = A_\sigma(f, L_\infty)$. Next using Lemma 10.2.2 for $N = 0$, we have

$$\limsup_{n \to \infty} E_n\big(f(\gamma_n^* \cdot), L_{\infty, W_1}(\mathbb{R})\big)$$
$$\leq \limsup_{n \to \infty} E_n\big(f(\gamma_n^* \cdot) - g(\gamma_n^* \cdot), L_{\infty, W_1}(\mathbb{R})\big) + \lim_{n \to \infty} E_n\big(g(\gamma_n^* \cdot), L_{\infty, W_1}(\mathbb{R})\big)$$
$$\leq A_\sigma(f, L_\infty).$$

Hence (10.2.1) follows.

Further let $P_n = \sum\limits_{k=0}^{n} d_{k,n} x^k$, $n = 2, 3, \ldots$, satisfy

(10.2.28) $\qquad \|f(\gamma_n^{**} \cdot) - P_n\|_{L_{\infty, W_1}[-a_n, a_n]} = E_n\big(f(\gamma_n^{**} \cdot), L_{\infty, W_1}[-a_n, a_n]\big).$

Hence

(10.2.29) $\qquad \|P_n\|_{L_{\infty, W_1}[-a_n, a_n]} \leq 2\|f\|_{L_\infty(\mathbb{R})}.$

Without loss of generality we may assume that

$$\lim_{n \to \infty} E_n\big(f(\gamma_n^{**} \cdot), L_{\infty, W_1}[-a_n, a_n]\big)$$

exists. Next using (10.2.29) and (10.2.5), we have ($k = 0, 1, \ldots, n$, $n = 2, 3, \ldots$)

$$|d_{k,n}| \leq \frac{3\sqrt{k+1}}{k!} \mu_n^k \|P_n\|_{L_{\infty, W_1}[-a_n, a_n]} \leq \frac{6\sqrt{k+1}}{k!} \mu_n^k \|f\|_{L_\infty(\mathbb{R})}.$$

Hence the coefficients of the polynomials $Q_n(x) := P_n(x/\gamma_n^{**}) = \sum\limits_{k=0}^{n} c_{k,n} x^k$ satisfy

$$|c_{k,n}| \leq \frac{6\sqrt{k+1}}{k!} \left(\frac{\mu_n \sigma}{\gamma_n^{**}}\right)^k \|f\|_{L_\infty(\mathbb{R})}, \qquad k = 0, 1, \ldots, n, \quad n = 2, 3, \ldots.$$

Since by (10.2.6), $\mu_n/\gamma_n^{**} = 1 + o(1)$, as $n \to \infty$, for every $\varepsilon \in (0, 1)$ there exists $n_0(\varepsilon)$ such that

$$|c_{k,n}| \leq \frac{6\sqrt{k+1}}{k!} \sigma^k (1+\varepsilon)^k \|f\|_{L_\infty(\mathbb{R})}, \quad k = 0, 1, \ldots, n, \quad n = n_0, n_0+1, \ldots.$$

Applying Proposition 5.1.2(a) to the sequence $\{Q_n\}_{n=n_0}^\infty$, we can find a subsequence $\{Q_{n_s}\}_{s=1}^\infty$ and $g_0 \in B_{\sigma(1+\varepsilon)}$ such that

(10.2.30) $\qquad \lim_{s \to \infty} Q_{n_s}(z) = g_0(z)$

uniformly in each interval $[-M, M]$.

Further, for a fixed numbers $M > 0$, $\delta \in (0, 1)$ and for n large enough, we have

$$\min_{x \in [-M, M]} \exp(-|x|/\gamma_n^{**}) = \exp(-M/\gamma_n^{**}) \geq 1 - \delta,$$

whence it follows that

$$E_n\big(f(\gamma_n^{**} \cdot), L_{\infty, W_1}[-a_n, a_n]\big) = \max_{y \in [-a_n \gamma_n^{**}, a_n \gamma_n^{**}]} |f(y) - Q_n(y)|$$

(10.2.31) $\qquad \times \exp[-|y|/\gamma_n^{**}] \geq (1-\delta) \max\limits_{y \in [-M, M]} |f(y) - Q_n(y)|.$

Taking into account (10.2.30), we obtain from (10.2.31)

(10.2.32) $\qquad \liminf\limits_{n \to \infty} E_n\big(f(\gamma_n^{**} \cdot), L_{\infty, W_1}[-a_n, a_n]\big) \geq (1-\delta) \|f - g_0\|_{L_\infty[-M, M]}.$

Letting $M \to \infty$ and then $\delta \to 0$ in (10.2.32), we have
$$\text{(10.2.33)} \qquad \liminf_{n\to\infty} E_n\big(f(\gamma_n^{**}\cdot), L_{\infty,W_1}[-a_n, a_n]\big) \geq A_{\sigma(1+\varepsilon)}(f, L_\infty).$$
Finally letting $\varepsilon \to 0$ in (10.2.33) and using Proposition 5.2.3(a), we arrive at (10.2.2). Thus Theorem 10.2.1 is established for $f \in L_\infty(\mathbb{R})$.

Let now f satisfy (2.1.4) and (2.1.5). Then by Proposition 5.2.1, there exists $g_0 \in B_{\sigma_0}$ such that
$$A_{\sigma_0}(f, L_\infty) = \|f - g_0\|_{L_\infty(\mathbb{R})} < \infty.$$
Hence a function $F := f - g_0 \in L_\infty(\mathbb{R})$, and for $\sigma \geq \sigma_0$ we have
$$\text{(10.2.34)} \qquad A_\sigma(f, L_\infty) = A_\sigma(F, L_\infty).$$
It is easy to see that $|g_0(\gamma_n^* x)| \leq C\gamma_n^{*N}(1+|x|)^N$, $x \in \mathbb{R}$, and by Lemma 10.2.2, the relation
$$\text{(10.2.35)} \qquad \lim_{n\to\infty} E_n\big(g_0(\gamma_n^*\cdot), L_{\infty,W_1}(\mathbb{R})\big) = 0$$
holds for every $\varepsilon \in (0,1)$ and $\sigma \geq \sigma_0$. Moreover, using Lemma 10.2.2 for $g_0(\gamma_n^{**}\cdot) \in B_{\frac{2(1-\varepsilon)\log n}{3\pi}}$, where $\varepsilon = 1 - 3\sigma_0/\sigma$, we have for $\sigma > 3\sigma_0$
$$\text{(10.2.36)} \qquad \lim_{n\to\infty} E_n\big(g_0(\gamma_n^{**}\cdot), L_{\infty,W_1}(\mathbb{R})\big) = 0.$$
Next applying (10.2.2) to $F \in L_\infty(\mathbb{R})$ and using (10.2.34) and (10.2.36), we obtain for $\sigma > 3\sigma_0$
$$\begin{aligned} A_\sigma(f, L_\infty) &= A_\sigma(F, L_\infty) \leq \liminf_{n\to\infty} E_n\big(F(\gamma_n^{**}\cdot), L_{\infty,W_1}[-a_n, a_n]\big) \\ &\leq \liminf_{n\to\infty} E_n\big(f(\gamma_n^{**}\cdot), L_{\infty,W_1}[-a_n, a_n]\big) \\ &\quad + \lim_{n\to\infty} E_n\big(g_0(\gamma_n^{**}\cdot), L_{\infty,W_1}(\mathbb{R})\big) \\ &= \liminf_{n\to\infty} E_n\big(f(\gamma_n^{**}\cdot), L_{\infty,W_1}[-a_n, a_n]\big). \end{aligned}$$
This proves Theorem 10.2.1(b).

Finally applying (10.2.1) to $F \in L_\infty(\mathbb{R})$ and using (10.2.34) and (10.2.35), we have for $\sigma \geq \sigma_0$
$$\begin{aligned} \limsup_{n\to\infty} E_n\big(f(\gamma_n^*\cdot), L_{\infty,W_1}(\mathbb{R})\big) &\leq \limsup_{n\to\infty} E_n\big(F(\gamma_n^*\cdot), L_{\infty,W_1}(\mathbb{R})\big) \\ &\quad + \lim_{n\to\infty} E_n\big(g_0(\gamma_n^*\cdot), L_{\infty,W_1}(\mathbb{R})\big) \leq A_\sigma(F, L_\infty) = A_\sigma(f, L_\infty). \end{aligned}$$
Hence Theorem 10.2.1(a) follows. ∎

The following result is an immediate consequence of Theorem 10.2.1:

COROLLARY 10.2.5. *For any $\lambda > 0$,*
$$\begin{aligned} (\pi/2)^\lambda B_{\lambda,\infty} &\leq \liminf_{n\to\infty}(\log n)^\lambda E_n\big(f_\lambda, L_{\infty,W_1}[-\pi n/2, \pi n/2]\big) \\ &\leq \limsup_{n\to\infty}(\log n)^\lambda E_n\big(f_\lambda, L_{\infty,W_1}(\mathbb{R})\big) \leq (3\pi/2)^\lambda B_{\lambda,\infty}. \end{aligned}$$
In particular, for $\lambda = 1$,
$$\text{(10.2.37)} \qquad \begin{aligned} 0.28(\pi/2) &\leq \liminf_{n\to\infty} \log n \, E_n\big(f_1, L_{\infty,W_1}[-\pi n/2, \pi n/2]\big) \\ &\leq \limsup_{n\to\infty} \log n \, E_n\big(f_1, L_{\infty,W_1}(\mathbb{R})\big) \leq 0.85(\pi/2). \end{aligned}$$

Remarks

(a) It follows from (8.1.10) for $k = 1$ that
$$E_n(f_1, L_{\infty, W_1}(\mathbb{R})) < (\pi/2)/\log(2n+1), \qquad n = 1, 2, \ldots.$$
This Bernstein's estimate immediately implies the following relation:
(10.2.38) $$\limsup_{n \to \infty} \log n \, E_n(f_1, L_{\infty, W_1}(\mathbb{R})) \leq \pi/2.$$
Thus the right inequality in (10.2.37) is a slight improvement of (10.2.38).

(b) We conjecture that in Theorem 10.2.1(a) the constant γ_n^* can be replaced by $\gamma_n^{**}(1 - \varepsilon)$.

10.3. $W(x) = \exp(-|x|^\alpha)$, $0 < \alpha < 1$

For an exponential weight $W(x) = \exp(-Q(x))$, where Q is a positive and continuous function on $\mathbb{R}$ with $\lim_{|x| \to \infty} Q(x)/\log x = \infty$, we consider a subspace $L_{\infty, W}^0(\mathbb{R})$ of all $f \in L_{\infty, W}(\mathbb{R})$ satisfying the condition
$$\lim_{|x| \to \infty} f(x) W(x) = 0.$$

Next, let B_0 be the class of all entire functions of minimal exponential type, i.e., the class of all entire functions g, satisfying the condition: for every $\varepsilon > 0$ there exists $C = C(g, \varepsilon)$ such that
(10.3.1) $$|g(z)| \leq C e^{\varepsilon |z|}, \qquad z \in \mathbb{C}.$$

It is easy to verify that the class $\mathcal{P}$ of all polynomials is a subset of $L_{\infty, W}^0 \cap B_0$. The closure of $\mathcal{P}$ in the metric of $L_{\infty, W}(\mathbb{R})$ is denoted by $\mathcal{P}[W]$.

It is well-known [3, 59, 88, 103] that the condition $\int_{-\infty}^{\infty} \frac{Q(x)}{1+x^2} dx = \infty$ is necessary for $\mathcal{P}[W] = L_{\infty, W}^0(\mathbb{R})$ (cf. Theorem 1.1.1). The problem of finding a description for $\mathcal{P}[W]$ in the case
(10.3.2) $$\int_{-\infty}^{\infty} \frac{Q(x)}{1+x^2} dx < \infty$$
was discussed in [3, 4, 59, 66, 88, 91]. In particular, Mergelyan [4, 59, 88] showed that (10.3.2) implies
(10.3.3) $$\mathcal{P}[W] \subset B_0 \cap L_{\infty, W}^0(\mathbb{R}).$$

Using Mergelyan's method, Kroó, Szabados, and Varga [66, Theorem 1] (see also [91, p. 331]) proved that under some mild conditions, $\mathcal{P}[W]$ is a subset of a class of entire functions, satisfying a more delicate condition than (10.3.1). A partial converse to Theorem 10.3.1 was given in [66, Theorem 5] (see also [91, p. 331]).

Let us consider the class $E_\alpha(\sigma)$ of the restrictions to $\mathbb{R}$ of all entire functions g of order $\alpha \in (0, 1)$ and type σ, satisfying
(10.3.4) $$|g(z)| \leq C(g) \exp(\sigma |z|^\alpha), \qquad z \in \mathbb{C}.$$

A typical example of a weight satisfying (10.3.2) is $W_\alpha(x) = \exp(-|x|^\alpha)$, $0 < \alpha < 1$. For this weight, the results from [66] imply the following inclusions
(10.3.5) $$E_\alpha(1 - \varepsilon) \subseteq \mathcal{P}[W_\alpha] \subseteq E_\alpha(5 \times 2^{\alpha - 1}(1 - \alpha)^{-1}).$$

In the following theorem, we characterize $\mathcal{P}[W_\alpha]$ in terms of the classes $E_\alpha(\sigma) \cap L^0_{\infty, W_\alpha}(\mathbb{R})$.

THEOREM 10.3.1. (a) *The following inclusions hold:*

(10.3.6) $\quad E_\alpha(\sigma_{1,\varepsilon}) \cap L^0_{\infty, W_\alpha}(\mathbb{R}) \subseteq \mathcal{P}[W_\alpha] \subseteq E_\alpha(\sigma_2) \cap L^0_{\infty, W_\alpha}(\mathbb{R}), \quad 0 < \alpha < 1,$

where

$$\sigma_{1,\varepsilon} = \sigma_{1,\varepsilon}(\alpha) := \sigma_0(\alpha)/(1+\varepsilon), \quad \sigma_2 = \sigma_2(\alpha) := \bigl(\cos(\alpha\pi/2)\bigr)^{-1}.$$

Here, ε is a positive number and

$$\sigma_0(\alpha) = \sigma_0 := B_\alpha \left(\frac{1 + r_0^{*2}}{2r_0^*}\right)^{2-\alpha} \log \frac{1}{r_0^*},$$

where B_α is given at (10.1.2) and $r_0^* = r_0^*(\alpha) \in (0,1)$ is the unique solution to the equation

(10.3.7) $\quad \dfrac{8r^2}{\alpha(1-r^2)^2} \left(F\left(-\dfrac{\alpha}{2}, 1, \dfrac{\alpha}{2}+1, r^2\right) - \dfrac{1}{2} \right) + \log r = 0.$

(b) *Moreover, if $g \in E_\alpha(\sigma_{1,\varepsilon}) \cap L^0_{\infty, W_\alpha}(\mathbb{R})$, $\varepsilon > 0$, then*

(10.3.8) $\quad E_n\bigl(g, L_{\infty, W_\alpha}(\mathbb{R})\bigr) \leq C n^\gamma \exp(-C_1 \varepsilon n) + \|g\|_{L_{\infty, W_\alpha}(|x| \geq a_n)}.$

PROOF. This consists of 4 steps. Using Mergelyan's method [4, 59, 66, 88], we first establish the right inclusion in (10.3.6) (Steps 1 and 2). Then applying the approximation techniques developed in Chapter 6, we prove statement (b) of the theorem which implies the left inclusion in (10.3.6) (Steps 3 and 4).

Step 1. We first prove the following polynomial inequality:

(10.3.9) $\quad |P(z)| \leq \exp\bigl(|z|^\alpha / \cos(\alpha\pi/2)\bigr) \|P\|_{L_{\infty, W_\alpha}(\mathbb{R})}, \quad z \in \mathbb{C}, \ 0 < \alpha < 1.$

This inequality is trivial for $z \in \mathbb{R}$. Let $z = x + iy \in \mathbb{C}$ with $y \neq 0$. Then by Theorem 6.5.4 in [19], we have

$$\begin{aligned}
\log |P(z)| &\leq \frac{|y|}{\pi} \int_{-\infty}^\infty \frac{\log |P(t)|}{(t-x)^2 + y^2} dt \\
&\leq \frac{|y|}{\pi} \int_{-\infty}^\infty \frac{|t|^\alpha + \log \|P\|_{L_{\infty, W_\alpha}(\mathbb{R})}}{t^2 - 2tx + |z|^2} dt = \log \|P\|_{L_{\infty, W_\alpha}(\mathbb{R})} \\
&\quad + \frac{|y|}{\pi} \biggl(\int_0^\infty \frac{t^\alpha}{t^2 - 2t|z|\cos\gamma + |z|^2} dt \\
&\quad + \int_0^\infty \frac{t^\alpha}{t^2 + 2t|z|\cos\gamma + |z|^2} dt \biggr),
\end{aligned}$$

(10.3.10)

where $\cos\gamma := |x|/|z|$ and $\gamma \in (0, \pi/2]$. Evaluating the Euler integrals in the right-hand side of (10.3.10) (see [54]), we get

$$\begin{aligned}
\log |P(z)| &\leq \log \|P\|_{L_{\infty, W_\alpha}(\mathbb{R})} + \frac{|z|^\alpha}{\sin(\alpha\pi)} \bigl(\sin(\alpha\gamma) + \sin(\alpha(\pi - \gamma))\bigr) \\
&\leq \log \|P\|_{L_{\infty, W_\alpha}(\mathbb{R})} + \frac{|z|^\alpha}{\cos(\alpha\pi/2)}.
\end{aligned}$$

This yields (10.3.9).

Step 2. Next we prove the right inclusion in (10.3.6). Let $g \in \mathcal{P}[W_\alpha]$, i.e., $g \in L^0_{\infty,W_\alpha}(\mathbb{R})$ and there exists a subsequence of $P_n \in \mathcal{P}_n$, $n = 1, 2, \ldots$, such that

$$(10.3.11) \quad \lim_{n\to\infty} E_n\big(g, L_{\infty,W_\alpha}(\mathbb{R})\big) = \lim_{n\to\infty} \|g - P_n\|_{L_{\infty,W_\alpha}(\mathbb{R})} = 0.$$

Hence for $n = 1, 2, \ldots$,

$$\|P_n\|_{L_{\infty,W_\alpha}(\mathbb{R})} \leq E_n\big(g, L_{\infty,W_\alpha}(\mathbb{R})\big) + \|g\|_{L_{\infty,W_\alpha}(\mathbb{R})} \leq 2\|g\|_{L_{\infty,W_\alpha}(\mathbb{R})}.$$

Then using (10.3.9), we get

$$(10.3.12) \quad |P_n(z)| \leq 2\|g\|_{L_{\infty,W_\alpha}(\mathbb{R})} \exp\big(|z|^\alpha / \cos(\alpha\pi/2)\big), \quad z \in \mathbb{C}.$$

Thus, all polynomials P_n, $n = 1, 2, \ldots$, are uniformly bounded on any disk in $\mathbb{C}$, and by the compactness theorems for analytic functions [115, Sections 5.2.1, 5.2.2], there exists a subsequence $\{n_k\}_{k=1}^\infty$ and an entire function g_0 such that $P_{n_k} \to g_0$ uniformly on any compact subset of $\mathbb{C}$. It follows from (10.3.12) that the restriction of g_0 to $\mathbb{R}$ belongs to $E_\alpha(\sigma_2)$. Also, (10.3.11) implies that $g_0 = g$. Therefore $g \in E_\alpha(\sigma_2) \cap L^0_{\infty,W_1}(\mathbb{R})$.

Step 3. To prove Theorem 10.3.1(b), we need some auxiliary functions.

We first study the function

$$(10.3.13) \quad G_{\alpha,\sigma}(r) := \sigma\left[\frac{a_n}{2}\left(r + \frac{1}{r}\right)\right]^\alpha - H(ir) + n\log r, \quad 0 < r < 1,\ \sigma > 0,\ 0 < \alpha < 1,$$

where a_n and $H(ir)$ are given in (10.1.3) and (10.1.5), respectively. It is easy to verify the formula

$$(10.3.14) \quad G'_{\alpha,\sigma}(r) = \frac{\alpha a_n^\alpha (1+r^2)^{\alpha-1}(1-r^2)}{2^\alpha r^{\alpha+1}}(h_2(r) - \sigma),$$

where

$$(10.3.15) \quad h_2(r) := \frac{2^{\alpha-1} r^{\alpha-1}}{\alpha a_n^{\alpha-1}(1+r^2)^{\alpha-2}(1-r^2)} h(r)$$

and h is given in (10.1.6).

Further, we note that h_2 is increasing in $(0,1)$. To show that, we denote $F(x) := F(-\alpha/2, 1, \alpha/2+1, x)$, $x \in (0,1)$, and consider

$$\frac{d}{dx}\left(\frac{F(x)-1/2}{1-x}\right) = \frac{F(x) - F(1) - F'(x)(x-1)}{(1-x)^2} = -\frac{x_1 - x}{1-x} F''(x_2)$$

$$(10.3.16) \quad = \frac{2\alpha(2-\alpha)(x_1-x)}{(\alpha+2)(\alpha+4)(1-x)} F(2-\alpha/2, 3, \alpha/2+3, x_2) > 0,$$

where $x < x_2 < x_1 < 1$. Then it follows from (10.1.6), (10.3.15), and (10.3.16) that $h(r)/r$ and so $h_2(r)$ are increasing in $r \in (0,1)$. Since $\lim_{r\to 0+} h_2(r) = 0$ and $\lim_{r\to 1-} h_2(r) = \infty$, the equation

$$h_2(r) = \sigma$$

has the unique continuously differentiable solution $r(\sigma)$ for every $\sigma > 0$.

Taking account of (10.3.14), we conclude that $\inf_{\rho\in(0,1)} G_{\alpha,\sigma}(\rho)$ attains at $\rho = r(\sigma)$, i.e.,

$$\begin{aligned}
\phi_2(r) &= \phi_2(r(\sigma)) := \inf_{\rho\in(0,1)} G_{\alpha,\sigma}(\rho) = G_{\alpha,\sigma}(r(\sigma)) \\
&= G_{\alpha,h_2(r)}(r) = \frac{a_n h(r)(1+r^2)^2}{2\alpha(1-r^2)r} - H(ir) + n\log r \\
&= n\left[\frac{8r^2}{\alpha(1-r^2)^2}\left(F\left(-\frac{\alpha}{2},1,\frac{\alpha}{2}+1,r^2\right) - \frac{1}{2}\right) + \log r\right].
\end{aligned}$$
(10.3.17)

Next by (10.3.16), ϕ_2 is increasing in $(0,1)$ and $\phi_2(0) = -\infty$, $\phi_2(1) = \infty$. Hence equation (10.3.7) has the unique solution $r_0^* = r_0^*(\alpha) \in (0,1)$ and for all $\varepsilon > 0$,

$$\inf_{\rho\in(0,r_0^*)} G_{\alpha,\sigma_{1,\varepsilon}}(\rho) = \phi_2(r(\sigma_{1,\varepsilon})) \leq -C(r(\sigma_0) - r(\sigma_{1,\varepsilon}))n.$$
(10.3.18)

Note that by (10.3.15), (10.1.6), and (10.1.3),

$$h_2(r) = \frac{2^{\alpha+1} B_\alpha r^\alpha}{\alpha(1-r^2)^2(1+r^2)^{\alpha-2}}\left(F\left(-\frac{\alpha}{2},1,\frac{\alpha}{2}+1,r^2\right) - \frac{1}{2}\right).$$
(10.3.19)

Consequently h_2 is independent of n and

$$r(\sigma_0) - r(\sigma_{1,\varepsilon}) = h_2^{-1}(\sigma_0) - h_2^{-1}(\sigma_{1,\varepsilon}) \geq C_1(\sigma_0 - \sigma_{1,\varepsilon}).$$
(10.3.20)

Then combining (10.3.18) with (10.3.20), we have

$$\inf_{\rho\in(0,r_0^*)} G_{\alpha,\sigma_{1,\varepsilon}}(\rho) \leq -C(\sigma_0 - \sigma_{1,\varepsilon})n \leq -C_2\varepsilon n.$$
(10.3.21)

We recall that $\sigma_0 = h_2(r_0^*)$, $\sigma_{1,\varepsilon} = \sigma_0/(1+\varepsilon)$, and $r(\sigma_0) = r_0^*$. Since $\phi_2(r_0^*) = 0$, (10.3.17) implies that r_0^* is the solution to (10.3.7). Moreover, by (10.3.7) and (10.3.19),

$$\begin{aligned}
\sigma_0 &= \frac{2^{\alpha+1} B_\alpha r_0^{*\alpha}}{\alpha(1-r_0^{*2})^2(1+r_0^{*2})^{\alpha-2}}\left(F\left(-\frac{\alpha}{2},1,\frac{\alpha}{2}+1,r_0^{*2}\right) - \frac{1}{2}\right) \\
&= B_\alpha\left(\frac{1+r_0^{*2}}{2r_0^*}\right)^{2-\alpha}\log\frac{1}{r_0^*}.
\end{aligned}$$

Step 4. Now we consider the approximation of $g \in E_\alpha(\sigma_{1,\varepsilon}) \cap L^0_{\infty,W_\alpha}(\mathbb{R})$ by the Lagrange interpolation polynomials $P_{n-1} \in \mathcal{P}_{n-1}$ to g at the zeros $\{x^{(q)}_{k,n}\}_{k=1}^n$ of the extremal polynomials $T_{n,q}(W_\alpha,\cdot)$ with $1 < q < \infty$ and $0 < \alpha < 1$.

Note that g is an entire function of exponential type τ for any $\tau > 0$, so that we can use estimates of $\|g - P_{n-1}\|_{L_{\infty,W_\alpha}(\mathbb{R})}$ established in Chapter 6.

Similarly to (6.0.17) we get

$$\begin{aligned}
E_n(g, L_{\infty,W_\alpha}(\mathbb{R})) &\leq \|g - P_{n-1}\|_{L_{\infty,W_\alpha}(\mathbb{R})} \leq \|g - P_{n-1}\|_{L_{\infty,W_\alpha}(I_n)} \\
&\quad + \|g - P_{n-1}\|_{L_{\infty,W_\alpha}(\mathbb{R}\setminus I_n)} = J_1 + J_2,
\end{aligned}$$
(10.3.22)

where $I_n = [-a_n(1+\delta_n), a_n(1+\delta_n)]$, and the numbers

$$\delta_n = \delta_0 := \min\bigl(\varepsilon^{2/3}, (1/r_0^*(\alpha) - 1)^2/30, 1/2\bigr), \qquad n = 1, 2, \ldots,$$

satisfy condition (6.0.16).

Since $W_\alpha \in \mathcal{F}(C^3)$ and $g \in E_\alpha(\sigma_{1,\varepsilon}) \subseteq B_\tau$, $\tau > 0$, we apply Proposition 6.1.2 to g and obtain that for all $r \in \left(0, 1 - \frac{5\sqrt{\delta_0}}{1+5\sqrt{\delta_0}}\right)$, the following inequality holds:

$$
\begin{aligned}
J_1 &\leq Cn^\gamma r^{n-1} e^{-H(ir)} \max_{|w|=r} \left| g\left(\frac{a_n}{2}\left(w + \frac{1}{w}\right)\right) \right| \\
&\leq Cn^\gamma r^{-1} \exp\left[G_{\alpha,\sigma_{1,\varepsilon}}(r)\right],
\end{aligned}
\tag{10.3.23}
$$

where $G_{\alpha,\sigma}$ is defined in (10.3.13).

Next, $(0, r_0^*] \subseteq \left(0, 1 - \frac{5\sqrt{\delta_0}}{1+5\sqrt{\delta_0}}\right)$, therefore relations (10.3.21) and (10.3.23) imply

$$
\begin{aligned}
J_1 &\leq Cn^\gamma \left(1/h^{-1}(\sigma_{1,\varepsilon})\right) \exp\left(\inf_{\rho \in (0, r_0^*]} G_{\alpha,\sigma_{1,\varepsilon}}(\rho)\right) \\
&\leq C_1(\varepsilon) n^\gamma \exp(-C_2 \varepsilon n).
\end{aligned}
\tag{10.3.24}
$$

Further using Proposition 6.2.2, we get

$$
J_2 \leq Cn^{m_1} \exp(-C_1 \delta_0^{3/2} n)\left(J_1 + \|g\|_{L_\infty, W_\alpha(\mathbb{R})}\right) + \|g\|_{L_\infty, W_\alpha(|x| \geq a_n)}.
\tag{10.3.25}
$$

Since $\|g\|_{L_\infty, W_\alpha(\mathbb{R})} < \infty$ and $\delta_0 < \varepsilon^{2/3}$, inequalities (10.3.22), (10.3.24), and (10.3.25) yield (10.3.8).

Finally, we recall that $g \in L^0_{\infty, W_\alpha}(\mathbb{R})$, i.e.,

$$
\lim_{n \to \infty} \|g\|_{L_\infty, W_\alpha(|x| \geq a_n)} = 0.
\tag{10.3.26}
$$

Thus (10.3.8) and (10.3.26) imply that $g \in \mathcal{P}[W_\alpha]$. This completes the proof of Theorem 10.3.1. ∎

COROLLARY 10.3.2. *If $g \in E_\alpha(\sigma)$, where $0 < \sigma < \min(\sigma_0(\alpha), 1)$, then*

$$
\limsup_{n \to \infty} \left(E_n(g, L_{\infty, W_\alpha}(\mathbb{R}))\right)^{1/n} < 1.
\tag{10.3.27}
$$

PROOF. Since $\sigma \in (0, 1)$, we have by (10.1.3),

$$
\|g\|_{L_\infty, W_\alpha(|x| \geq a_n)} \leq C \exp[(\sigma - 1) a_n^\alpha] \leq C \exp(-C_1 n).
\tag{10.3.28}
$$

Then setting $\varepsilon = \sigma_0(\alpha)/\sigma - 1$, we see that $g \in E_\alpha(\sigma_{1,\varepsilon}) \cap L^0_{\infty, W_\alpha}(\mathbb{R})$, so that (10.3.27) follows from (10.3.8) and (10.3.28). ∎

Remarks

(a) The class $E_\alpha(\sigma) \cap L^0_{\infty, W_\alpha}(\mathbb{R})$ is not trivial for any $\sigma > 0$ and $0 < \alpha < 1$, since any entire function of order $\alpha_1 \in (0, \alpha)$ belongs to this class, for example

$$
\sum_{n=0}^\infty (\sigma z)^n / \Gamma(\alpha_1 h + 1) \in E_\alpha(\sigma) \cap L^0_{\infty, W_\alpha}(\mathbb{R}).
$$

Moreover, for any even function $g \in E_\alpha(\sigma)$, there exists $t_0 \in [0, 2\pi)$ such that $g(e^{it_0} z) \in E_\alpha(\sigma) \cap L^0_{\infty, W_\alpha}(\mathbb{R})$.

(b) Since $\sigma_2(\alpha) < (1-\alpha)^{-1}$, the right inclusion in (10.3.6) strengthens the right inclusion in (10.3.5). The comparison of the left inclusions in (10.3.5) and (10.3.6) is difficult because of the intricate definition of σ_0.

10.4. $W(x) = \exp(-|x|^\alpha)$, $\alpha \to \infty$

Here we discuss limit relations of different type.

THEOREM 10.4.1. *If $f \in L_{p,W_a}(\mathbb{R})$ for $0 < p \leq \infty$ and for some $a > 1$, then*

(10.4.1) $$\lim_{\alpha \to \infty} E_n(f, L_{p,W_\alpha}(\mathbb{R})) = E_n(f, L_p[-1,1]).$$

Moreover, let $P_{n,\alpha}(f, \cdot)$ and $P_{n,\infty}(f, \cdot)$ be best approximating polynomials to f of degree at most n in the metrics of $L_{p,W_\alpha}(\mathbb{R})$ and $L_p[-1,1]$, $1 \leq p \leq \infty$, respectively. Then for $p \in (1, \infty]$ we have

(10.4.2) $$\lim_{\alpha \to \infty} P_{n,\alpha}(f, x) = P_{n,\infty}(f, x)$$

uniformly on any compact set of $\mathbb{R}$. For $p = 1$ (10.4.2) holds if f is continuous on $[-1, 1]$.

PROOF. We first prove (10.4.1). For simplicity, we assume that $1 \leq p \leq \infty$. The proof for $0 < p < 1$ is similar.

For any $\varepsilon > 0$ there exists $\delta = \delta(\varepsilon, f, P_{n,\infty})$ such that

(10.4.3) $$\|f - P_{n,\infty}\|_{L_p[-\delta-1, 1+\delta]} < E_n(f, L_p[-1,1]) + \varepsilon.$$

Next, there exists $\alpha_0 = \alpha_0(\varepsilon, \delta, f, P_{n,\infty})$ such that for $\alpha > \max(a, \alpha_0)$,

(10.4.4) $$\begin{aligned} \|f - P_{n,\infty}\|_{L_{p,W_\alpha}(|x|\geq 1+\delta)} &\leq \max_{|x|\geq 1+\delta} \exp\bigl(-|x|^a(|x|^{\alpha-a}-1)\bigr) \\ &\quad \times \|f - P_{n,\infty}\|_{L_{p,W_a}(|x|\geq 1+\delta)} \\ &\leq 3\exp\bigl(-(1+\delta)^{\alpha-a}\bigr)\|f - P_{n,\infty}\|_{L_{p,W_a}(\mathbb{R})} < \varepsilon. \end{aligned}$$

Then using (10.4.3) and (10.4.4), we get for all α large enough

$$\begin{aligned} E_n(f, L_{p,W_\alpha}(\mathbb{R})) &\leq \|f - P_{n,\infty}\|_{L_{p,W_\alpha}[-1-\delta, 1+\delta]} \\ &\quad + \|f - P_{n,\infty}\|_{L_{p,W_\alpha}(|x|\geq 1+\delta)} \\ &\leq E_n(f, L_p[-1,1]) + 2\varepsilon. \end{aligned}$$

Hence

(10.4.5) $$\limsup_{\alpha \to \infty} E_n(f, L_{p,W_\alpha}(\mathbb{R})) \leq E_n(f, L_p[-1,1]).$$

Further for every $\delta \in (0, 1)$,

$$\begin{aligned} \liminf_{\alpha \to \infty} E_n(f, L_{p,W_\alpha}(\mathbb{R})) &\geq \liminf_{\alpha \to \infty} \exp\bigl(-(1-\delta)^\alpha\bigr)\|f - P_{n,\alpha}\|_{L_p[-1+\delta, 1-\delta]} \\ &\geq E_n(f, L_p[-1+\delta, 1-\delta]). \end{aligned}$$

These inequalities show that in order to prove the estimate

(10.4.6) $$\liminf_{\alpha \to \infty} E_n(f, L_{p,W_\alpha}(\mathbb{R})) \geq E_n(f, L_p[-1,1]),$$

it remains to establish the relation

(10.4.7) $$\lim_{\delta \to 0+} E_n(f, L_p[-1+\delta, 1-\delta]) = E_n(f, L_p[-1,1]).$$

Indeed, let $R_{n,\delta} \in \mathcal{P}_n$ satisfy the equality

(10.4.8) $$E_n(f, L_p[-1+\delta, 1-\delta]) = \|f - R_{n,\delta}\|_{L_p[-1+\delta, 1-\delta]}.$$

Without loss of generality we may assume that there exists $R_{n,0} \in \mathcal{P}_n$ such that

(10.4.9) $$\lim_{\delta \to 0} \|R_{n,\delta} - R_{n,0}\|_{L_p[-1,1]} = 0.$$

Next note that for any $\varepsilon > 0$ there exists $\delta_1 = \delta_1(\varepsilon, f, R_{n,0})$ such that for all $\delta \in (0, \delta_1)$

(10.4.10) $$\|f - R_{n,0}\|_{L_p[-1+\delta, 1-\delta]} > \|f - R_{n,0}\|_{L_p[-1,1]} - \varepsilon.$$

Using (10.4.8)–(10.4.10), we have that for all $\delta > 0$ small enough

$$\begin{aligned} E_n(f, L_p[-1+\delta, 1-\delta]) &\geq \|f - R_{n_0}\|_{L_p[-1+\delta, 1-\delta]} - \|R_{n,\delta} - R_{n,0}\|_{L_p[-1,1]} \\ &> \|f - R_{n,0}\|_{L_p[-1,1]} - 2\varepsilon \geq E_n(f, L_p[-1,1]) - 2\varepsilon. \end{aligned}$$

This yields (10.4.7). Therefore (10.4.6) holds. Finally, (10.4.5) and (10.4.6) imply (10.4.1).

To prove (10.4.2), we first note that if $1 < p \leq \infty$ or f is continuous on $[-1,1]$ and $p = 1$, then $P_{n,\infty}$ is the unique best approximating polynomial to f in $L_p[-1,1]$ (cf. [4, 25]).

Let $\{\alpha(k)\}_{k=1}^\infty$ be an increasing sequence of numbers from (a, ∞). Without loss of generality we may assume that there exists $R_n \in \mathcal{P}_n$ such that

(10.4.11) $$\lim_{k \to \infty} \|P_{n,\alpha(k)} - R_n\|_{L_\infty[-1,1]} = 0.$$

Then by (10.4.1) and (10.4.11), we have

$$\begin{aligned} E_n(f, L_p[-1,1]) &= \lim_{k \to \infty} E_n(f, L_{p,W_{\alpha(k)}}(\mathbb{R})) \\ &\geq \limsup_{k \to \infty} \|f - P_{n,\alpha(k)}\|_{L_{p,W_{\alpha(k)}}[-1,1]} \\ &\geq \limsup_{k \to \infty} \|f - R_n\|_{L_{p,W_{\alpha(k)}}[-1,1]} \\ &\quad - C \lim_{k \to \infty} \|P_{n,\alpha(k)} - R_n\|_{L_\infty[-1,1]} \end{aligned}$$

(10.4.12) $$= \limsup_{k \to \infty} \|f - R_n\|_{L_{p,W_{\alpha(k)}}[-1,1]}.$$

Next it is easy to show that

(10.4.13) $$\lim_{\alpha \to \infty} \|f - R_n\|_{L_{p,W_\alpha}[-1,1]} = \|f - R_n\|_{L_p[-1,1]}$$

(cf. the proof of (10.4.6) above). Then combining (10.4.12) with (10.4.13), we get $R_n = P_{n,\infty}$. Thus (10.4.11) yields (10.4.2). ∎

We apply Theorem 10.4.1 to the normalized extremal polynomials $P_{n,p}(W_\alpha, \cdot)$.

COROLLARY 10.4.2. *If $1 \leq p \leq \infty$, then*

$$\lim_{\alpha \to \infty} P_{n,p}(W_\alpha, x) = T_{n,p}^*(x)/\|T_{n,p}^*\|_{L_p[-1,1]}$$

uniformly on any compact set of $\mathbb{R}$. Here, $T_{n,p}^*(x) = x^n + \cdots$ is the polynomial that deviates least from zero in $L_p[-1,1]$.

Remarks

(a) The polynomials $T_{n,p}^*$ which deviate least from zero in $L_p[-1,1]$ are well known for $p = \infty, 1, 2$ [114, Section 2.9], so that the following relations can be easily deduced from Corollary 10.4.2:

$$\lim_{\alpha \to \infty} P_{n,\infty}(W_\alpha, x) = T_n(x),$$
$$\lim_{\alpha \to \infty} P_{n,1}(W_\alpha, x) = (1/2)U_n(x),$$
$$\lim_{\alpha \to \infty} P_{n,2}(W_\alpha, x) = \sqrt{(2n+1)/2}\, P_n(x),$$

where T_n and U_n are the Chebyshev polynomials of first and second kind, respectively, and P_n is the Legendre polynomial.

(b) It is not difficult to show that

(10.4.14) $$\lim_{\alpha \to \infty} a_n(Q_\alpha) = 1, \qquad \lim_{\alpha \to \infty} b_n(Q_\alpha) = n.$$

Then Theorem 10.4.1 shows that the non-weighted limit theorems given by (1.1.5) and (1.1.6) are the limit cases as $\alpha \to \infty$ of the corresponding results with the weight W_α (see Theorems 2.1.1 and 2.1.2). However, we cannot deduce (1.1.5) and (1.1.6) directly from Theorems 2.1.1, 2.1.2, and 10.4.1.

10.5. Examples of Erdös Weights

In this section, we consider a few examples of Erdös weights from $\mathcal{F}(C^2)$ on $\mathbb{R}$ and $(-1, 1)$ and discuss asymptotics for a_n and b_n as $n \to \infty$; more information can be found in [72].

$W(x) = \exp(-\exp_\ell(|x|^\alpha) + \exp_\ell(0)), \alpha > 1, I = \mathbb{R}$. We define the ℓth iterated exponential $\exp_\ell(x)$ and the ℓth iterated logarithm $\log_\ell(x)$ by the recurrence formulae

$$\exp_0(x) := x, \quad \exp_k(x) := \exp(\exp_{k-1}(x)), \quad k = 1, 2, \ldots,$$
$$\log_0(x) := x, \quad \log_k(x) := \log(\log_{k-1}(x)), \quad x > \exp_{k-1}(0),\ k = 1, 2, \ldots.$$

Let $I = \mathbb{R}$ and

$$Q(x) = Q_{\ell,\alpha}(x) := \exp_\ell(|x|^\alpha) - \exp_\ell(0), \quad x \in \mathbb{R},\ \alpha > 1.$$

It is clear that $Q_{\ell,\alpha}$ is increasing and convex in $(0, \infty)$ with $Q_{\ell,\alpha}(0) = 0$ and $Q_{\ell,\alpha}(\infty) = \infty$. Next for $\ell \geq 1$ and $x > 0$,

$$T(x) = \frac{\alpha x^\alpha \prod_{k=1}^{\ell} \exp(x^\alpha)}{\exp_\ell(x^\alpha) - \exp_\ell(0)}.$$

Then it is not difficult to verify by straightforward calculations that T is quasi-increasing for $x > 0$, $\lim_{x \to \infty} T(x) = \infty$ and

$$\Lambda := \inf_{x > 0} T(x) = \lim_{x \to 0} T(x) = \alpha.$$

Also, (1.4.6) holds for some $C > 0$. Thus $W \in \mathcal{F}(C^2)$ with $\Lambda = \alpha > 1$, and $W(x) = \exp(-Q_{\ell,\alpha}(x))$ is an Erdös weight.

It was shown in [72] that

$$a_n = \bigl(\log_\ell(n)\bigr)^{1/\alpha}(1 + o(1)), \qquad n \to \infty.$$

Hence by Proposition 3.2.2(c),

$$b_n = \frac{n}{(\log_\ell(n))^{1/\alpha}}(1 + o(1)), \qquad n \to \infty.$$

10.5. EXAMPLES OF ERDÖS WEIGHTS

$W(x) = \exp(-(1-x^2)^{-\alpha} + 1), \alpha > 0, I = (-1,1)$. Let $I = (-1,1)$ and
$$Q(x) = Q^{(\alpha)}(x) := (1-x^2)^{-\alpha} - 1, \qquad x \in (-1,1).$$
Then elementary calculations show that $W(x) = \exp(-Q^{(\alpha)}(x))$ is an Erdös weight from $\mathcal{F}(C^2)$ with $\Lambda = 2$.

To find a_n and b_n we note that

(10.5.1) $\qquad n = \dfrac{2a_n}{\pi} \displaystyle\int_0^1 \dfrac{x(Q^{(\alpha)})'(a_n x)}{\sqrt{1-x^2}}\, dx = \dfrac{4\alpha}{\pi}(V_n(\alpha+1) - V_n(\alpha)),$

where

$$V_n(\gamma) := \int_0^1 \frac{dx}{(1-(a_n x)^2)^\gamma \sqrt{1-x^2}}$$

(10.5.2) $\qquad = \begin{cases} (\pi/2)(1-a_n^2)^{-\gamma/2} P_{-\gamma}\left[\frac{1-a_n^2/2}{\sqrt{1-a_n^2}}\right], & \gamma \neq 1/2 \\ K(a_n), & \gamma = 1/2. \end{cases}$

Here, $P_{-\gamma}(z) = P_{\gamma-1}(z)$ is the Legendre function and $K(t)$ the elliptic integral of first kind.

Using the representation of $P_{-\gamma}$ and $P_{1-\gamma}$ in terms of hypergeometric functions $F(a,b,c,d,w)$ [54], we have for $\gamma \neq 1/2$ as $z \to \infty$

$$\begin{aligned}
P_{\gamma-1}(z) &= \left(\frac{1+z}{2}\right)^{\gamma-1} F\left(1-\gamma, 1-\gamma, 1, \frac{z-1}{z+1}\right) \\
&= \left(\frac{z}{2}\right)^{\gamma-1} F(1-\gamma, 1-\gamma, 1, 1)(1+o(1)) \\
&= \left(\frac{z}{2}\right)^{\gamma-1} \frac{\Gamma(2\gamma-1)}{(\Gamma(\gamma))^2}(1+o(1)), \quad \gamma > 1/2, \\
P_{-\gamma}(z) &= \left(\frac{1+z}{2}\right)^{-\gamma} F\left(\gamma, \gamma, 1, \frac{z-1}{z+1}\right) \\
&= \left(\frac{z}{2}\right)^{-\gamma} F(\gamma, \gamma, 1, 1)(1+o(1)) \\
&= \left(\frac{z}{2}\right)^{-\gamma} \frac{\Gamma(1-2\gamma)}{(\Gamma(1-\gamma))^2}(1+o(1)), \quad \gamma < 1/2.
\end{aligned}$$

Combining these formulae with (10.5.2), and taking account of the facts that

$$a_n \to 1, \quad \frac{1-a_n^2/2}{\sqrt{1-a_n^2}} \to \infty, \quad n \to \infty,$$

we get as $n \to \infty$

(10.5.3) $\quad V_n(\gamma) = \pi 2^{1-2\gamma} \dfrac{\Gamma(2\gamma-1)}{(\Gamma(\gamma))^2}(1-a_n^2)^{1/2-\gamma}(1+o(1)), \quad \gamma > 1/2,$

(10.5.4) $\quad V_n(\gamma) = \pi 2^{-1-2\gamma} \dfrac{\Gamma(1-2\gamma)}{(\Gamma(1-\gamma))^2}(1+o(1)), \quad \gamma < 1/2.$

Next using the asymptotic for the elliptic integral $K(a_n)$ [33, Eq. (13.8.(10))], we obtain

(10.5.5) $\qquad V_n(1/2) = \log\left(1/\sqrt{1-a_n^2}\right)(1+o(1)), \quad n \to \infty.$

Asymptotic formulae (10.5.3)–(10.5.5) show that $V_n(\alpha) = o(V_n(\alpha+1))$, $n \to \infty$, $\alpha > 0$. Thus (10.5.1) and (10.5.3) imply as $n \to \infty$

$$(10.5.6) \quad n = \frac{4\alpha}{\pi} V_n(\alpha+1)(1+o(1)) = 2^{2-2\alpha} \frac{\Gamma(2\alpha)}{\Gamma^2(\alpha)} (1-a_n^2)^{-(\alpha+1/2)}(1+o(1)).$$

Therefore,

$$(10.5.7) \quad a_n = 1 - C(\alpha) n^{-1/(\alpha+1/2)}(1+o(1)), \qquad n \to \infty,$$

where

$$(10.5.8) \quad C(\alpha) := \frac{1}{2} \left[\frac{2^{2-2\alpha}\Gamma(2\alpha)}{\Gamma^2(\alpha)} \right]^{1/(\alpha+1/2)}.$$

Note that the relation $1 - a_n \sim n^{-1/(\alpha+1/2)}$ was given in [72, p. 31].

Further using (3.2.4) and (10.5.1)–(10.5.8), we have

$$\begin{aligned}
b_n &= \frac{2}{\pi} \int_0^1 \frac{(Q^{(\alpha)})'(a_n x)}{x\sqrt{1-x^2}} \, dx = \frac{4\alpha a_n}{\pi} V_n(\alpha+1) = a_n \left(n + \frac{4\alpha}{\pi} V_n(\alpha) \right) \\
&= a_n n + a_n(1+o(1)) \times \begin{cases} 2C(\alpha) n^{\frac{\alpha-1/2}{\alpha+1/2}}, & \alpha > 1/2 \\ \frac{4\alpha}{\pi(2\alpha+1)} \log n, & \alpha = 1/2 \\ \alpha 2^{1-2\alpha} \frac{\Gamma(1-2\alpha)}{(\Gamma(1-\alpha))^2}, & 0 < \alpha < 1/2 \end{cases} \\
&= n + (1+o(x)) \times \begin{cases} C(\alpha) n^{\frac{\alpha-1/2}{\alpha+1/2}}, & \alpha > 1/2 \\ \frac{4\alpha}{\pi(2\alpha+1)} \log n, & \alpha = 1/2 \\ \alpha 2^{1-2\alpha} \frac{\Gamma(1-2\alpha)}{(\Gamma(1-\alpha))^2}, & 0 < \alpha < 1/2. \end{cases}
\end{aligned}$$

$W(x) = \exp\bigl(-\exp_\ell((1-x^2)^{-\alpha}) + \exp_\ell(1)\bigr), \alpha > 0, I = (-1,1)$. A more general example of an exponential weight on $I = (-1,1)$ is $W(x) = \exp(-Q^{(\ell,\alpha)}(x))$, where

$$Q^{(\ell,\alpha)}(x) := \exp_\ell\bigl((1-x^2)^{-\alpha}\bigr) - \exp_\ell(1), \qquad \alpha > 0, \ \ell \geq 0.$$

It is an Erdös weight from $\mathcal{F}(C^2)$ with $\Lambda = 2$ and for $\ell \geq 1$,

$$a_n = 1 - \frac{1}{2}\bigl(\log_\ell(n)\bigr)^{-1/\alpha}(1+o(1))$$

(see [72, p. 33]). The asymptotic for b_n follows from Proposition 3.2.2(c)

$$b_n = n(1+o(1)), \qquad n \to \infty.$$

APPENDIX A

Appendix. Negativity of a Kernel

A.1. Statement of the Main Result

Negativity of the kernel $F(r,t,\varphi)$ defined at (3.4.32), where $(r,\varphi) \in K_1, K_2$, plays an essential role in the proof of Theorem 3.4.1. In this appendix we prove the inequality

(A.1.1) $\qquad F(r,t,\varphi) \leq 0, \quad 0 \leq t < \pi/2, \quad 0 \leq r \leq 1, \quad (r,\varphi) \in K_i, \quad i = 1,2.$

Let

(A.1.2) $\qquad F_i(r,t) := \sin t K\bigl(r,t,\varphi_i(r)\bigr) + 2\cos t \log r + f_i(r)/\cos t, \quad i = 1,2,$

where
(A.1.3)
$$K(r,t,\varphi) := \begin{cases} \arctan\left(\dfrac{1-r^2}{1+r^2}\cot(t+\varphi)\right) + \arctan\left(\dfrac{1-r^2}{1+r^2}\cot(t-\varphi)\right), \\ \qquad 0 \leq t < \varphi \leq \pi/2, \\ \arctan\left(\dfrac{1-r^2}{1+r^2}\cot(t+\varphi)\right) + \arctan\left(\dfrac{1-r^2}{1+r^2}\cot(t-\varphi)\right) - \pi, \\ \qquad \pi/2 \geq t > \varphi \geq 0, \\ \arctan\left(\dfrac{1-r^2}{1+r^2}\cot(2\varphi)\right) - \pi/2, \\ \qquad 0 \leq t = \varphi \leq \pi/2, \end{cases}$$

and

(A.1.4) $\qquad \varphi_1 = \varphi_1(r) := \arctan\left(\dfrac{2(1+r^2)}{1-r^2}\right), \quad f_1(r) := \left(\dfrac{1}{r} - r\right)\sin\varphi_1,$

(A.1.5) $\qquad \varphi_2 = \varphi_2(r) := \arctan\left(\dfrac{1+r^2}{2(1-r^2)}\right), \quad f_2(r) := \left(\dfrac{1}{r} + r\right)\cos\varphi_2.$

It is easy to see that (A.1.1) is equivalent to the following result:

PROPOSITION A.1.1. *For $i = 1, 2$, $0 \leq t < \pi/2$, and $0 \leq r < 1$, the inequalities*

(A.1.6) $\qquad\qquad\qquad\qquad F_i(r,t) \geq 0$

hold.

Despite the fact that $F_i(r,t)$ are combinations of elementary functions, the proof of Proposition A.1.1 is fairly long and technical. Since $F_1(r,t)$ is decreasing in r, a fairly short proof of the inequality $F_1(r,t) \geq 0$ is possible. Nevertheless, $F_2(r,t)$ has an irregular behavior, and some steps of the proof of the estimate $F_2(r,t) \geq 0$ are based on numerical computations.

We use certain *Mathematica* commands and packages firstly for evaluation of some elementary functions (in particular, $F_2(r,t)$, some polynomials, and others), and secondly for more sophisticated procedures such as finding $\min_{\Omega} F_2(r,t)$, where Ω is a finite set in $\mathbb{R}^2$, solving polynomial equations with integral coefficients, and counting the number of zeros of polynomials on a closed interval. We remark that all calculations were carried out to 30 significant digits.

A.2. Some Technical Results

To prove Proposition A.1.1, we need some technical estimates and also certain properties of F_i, $i=1,2$. We first establish some estimates of the form

(A.2.1) $$H(P,Q,R,r) := P(r) + Q(r)\sqrt{R(r)} \gtrless 0.$$

LEMMA A.2.1. *The following statements hold:*

(a) *If*
$$P_1(r) := 3 + 10r^2 + 3r^4, \quad Q_1(r) := -4r, \quad R_1(r) := 5 + 6r^2 + 5r^4,$$
then $H(P_1, Q_1, R_1, r) > 0$, $r \in [0,1)$.

(b) *If*
$$P_2(r) := -5 - 12r^2 - 30r^4 - 12r^6 - 5r^8,$$
$$Q_2(r) := 5r + 6r^3 + 5r^5, \quad R_2(r) := 5 + 6r^2 + 5r^4,$$
then $H(P_2, Q_2, R_2, r) \leq 0$, $r \in [0,1)$.

(c) *If*
$$P_3(r) := -20r + 24r^3 - 20r^5, \quad Q_3(r) := -3 + 10r^2 - 3r^4,$$
$$R_3(r) := 5 - 6r^2 + 5r^4,$$
then $H(P_3, Q_3, R_3, r) < 0$, $r \in [0,1]$.

(d) *If*
$$P_4(r) := 2560r^3, \quad Q_4(r) := 125 - 1174r^2 + 125r^4, \quad R_4(r) := 5 - 6r^2 + 5r^4,$$
then $H(P_4, Q_4, R_4) > 0$, $r \in [0,1]$.

(e) *If*
$$\begin{aligned}P_5(r) := {}& 2070000r - 42972000r^3 + 445217600r^5 - 2828739360r^7 \\ & + 11286670784r^9 - 32325190240r^{11} + 66658981056r^{13} \\ & - 103800366880r^{15} + 119270365344r^{17} - 103800366880r^{19} \\ & + 66658981056r^{21} - 32325190240r^{23} + 11286670784r^{25} \\ & - 2828739360r^{27} + 445217600r^{29} - 42972000r^{31} + 2070000r^{33},\end{aligned}$$
$$\begin{aligned}Q_5(r) := {}& 84375 + 846300r^2 - 30908240r^4 + 264267188r^6 \\ & - 1094557420r^8 + 3228108732r^{10} - 6366084080r^{12} \\ & + 9905318068r^{14} - 10833338070r^{16} + 9905318068r^{18} \\ & - 6366084080r^{20} + 3228108732r^{22} - 1094557420r^{24} \\ & + 264267188r^{26} - 30908240r^{28} + 846300r^{30} + 84375r^{32},\end{aligned}$$
$$R_5(r) := 5 - 6r^2 + 5r^4,$$
then $H(P_5, Q_5, R_5, r) < 0$, $r \in [0,1]$.

(f) *If*

$$
\begin{aligned}
P_6(r) := \ & -60558280209326407225 + 179814409592652270840 r^2 \\
& - 737471267926740517208 r^4 + 670162112460143936232 r^6 \\
& + 699148924368627358724 r^8 - 405830783803379792520 r^{10} \\
& + 1842458643546247118744 r^{12} - 887668079536219076952 r^{14} \\
& - 1167915724852400978 1270 r^{16} - 887668079536219076952 r^{18} \\
& + 1842458643546247118744 r^{20} - 405830783803379792520 r^{22} \\
& + 699148924368627358724 r^{24} + 670162112460143936232 r^{26} \\
& - 737471267926740517208 r^{28} + 179814409592652270840 r^{30} \\
& - 60558280209326407225 r^{32}, \\
Q_6(r) := \ & 53986603713392000425 r - 144934559836408003770 r^3 \\
& + 338158462921656974787 r^5 + 156624752035969120380 r^7 \\
& - 1160523642013405276991 r^9 + 1032563148831960252570 r^{11} \\
& + 768378575378356301779 r^{13} - 2088506682063042738360 r^{15} \\
& + 768378575378356301779 r^{17} + 1032563148831960252570 r^{19} \\
& - 1160523642013405276991 r^{21} + 156624752035969120380 r^{23} \\
& + 338158462921656974787 r^{25} - 144934559836408003770 r^{27} \\
& + 53986603713392000425 r^{29}, \\
R_6(r) := \ & 5 - 6 r^2 + 5 r^4,
\end{aligned}
$$

then $H(P_6, Q_6, R_6, r) < 0$. $r \in [0, 1]$.

PROOF. The proofs of all statements are based on analysis of the following polynomials with integral coefficients:

$$G_i(r) = G(P_i, Q_i, R_i, r) := P_i^2(r) - R_i(r) Q_i^2(r),$$

whose zeros in an interval J include all zeros of $H(P_i, Q_i, R_i, r)$ in J, $1 \leq i \leq 6$. In statements (a) and (b), G_i do not have zeros in $[0, 1)$, $i = 1, 2$.

(a) This statement immediately follows from the identity

$$G(P_1, Q_1, R_1, r) = (1 - r^2)^2 (9 - 2 r^2 + 9 r^4).$$

(b) It is easy to verify that

$$G(P_2, Q_2, R_2, r) = (1 - r^2)^2 (25 + 45 r^2 + 59 r^4 - 2 r^6 + 59 r^8 + 45 r^{10} + 25 r^{12}).$$

Hence $H(P_2, Q_2, R_2, r) < 0$, $r \in [0, 1)$.

In statements (c), (d), and (e), G_i has one or more zeros on $[0, 1]$. It turns out that for a zero ρ of G_i, $H(P_i, Q_i, R_i, \rho) \neq 0$, $3 \leq i \leq 5$. To show that, we first find an approximation ρ^* to ρ by using the *Mathematica* command N[Solve[$G_i[r]$ == 0], 28] that gives all zeros of the polynomial G_i with the precision 10^{-27}, $3 \leq i \leq 5$.

Next, we estimate the computational error Δ_i by the inequality

$$\Delta_i := \sup_{\substack{r,r^* \in [0,1] \\ |r-r^*| \leq 10^{-27}}} \left|H(P_i, Q_i, R_i, r) - H(P_i, Q_i, R_i, r^*)\right|$$

(A.2.2)
$$\leq 10^{-27}\left(\deg(P_i)\|P_i\|_{\ell_1} + \deg(Q_i)\|Q_i\|_{\ell_1}\sqrt{\|R_i\|_{\ell_1}} + \frac{\deg(R_i)\|R_i\|_{\ell_1}\|Q_i\|_{\ell_1}}{2\min_{r\in[0,1]}\sqrt{R_i(r)}}\right), \quad 3 \leq i \leq 5,$$

where $n = \deg(S)$ is the degree of a polynomial $S = \sum_{k=0}^{n} c_k x^k$ and $\|S\|_{\ell_1} := \sum_{k=1}^{n} |c_k|$.
Finally, we show that $H(P_i, Q_i, R_i, \rho) \neq 0$ by computing $H(P_i, Q_i, R_i, \rho^*)$ and taking account of (A.2.2).

(c) The only zero of $G(P_3, Q_3, R_3, r)$ on $[0,1]$ is

$$\rho_1 \approx \rho_1^* = 0.267949192431122706472553658.$$

This also follows from the identity

$$G(P_3, Q_3, R_3, r) = -(5 - 6r^2 + 5r^4)(1 - 4r + r^2)(1 + 4r + r^2)(9 - 14r^2 + 9r^4),$$

since $\rho_1 = 2 - \sqrt{3}$ is the only solution on $[0,1]$ to the equation $1 - 4r + r^2 = 0$.

Next, $H(P_3, Q_3, R_3, \rho_1^*) = -9.84\ldots$ and $\Delta_3 < 10^{-24}$. Therefore $H(P_3, Q_3, R_3, r) < 0$, $r \in [0,1]$.

(d) By a straight calculation, we have

$$\begin{aligned}G(P_4, Q_4, R_4, r) = & -78125 + 1561250r^2 - 8886755r^4 \\ & + 17945756r^6 - 8886755r^8 + 1561250r^{10} - 78125r^{12}.\end{aligned}$$

This polynomial has the unique zero

$$\rho_2 \approx \rho_2^* = 0.287766394156836431913237365$$

on $[0,1]$ and $H(P_4, Q_4, R_4, \rho_2^*) = 122.008\ldots$ with $\Delta_4 < 10^{-22}$. Thus $H(P_4, Q_4, R_4, r) > 0$, $r \in [0,1]$.

(e) The polynomial $G(P_5, Q_5, R_5, r)$ has the following 3 zeros on $[0,1]$:

$$\begin{aligned}\rho_3 &\approx \rho_3^* = 0.199829711896443536552697543, \\ \rho_4 &\approx \rho_4^* = 0.202804779877365821009692671, \\ \rho_5 &\approx \rho_5^* = 0.267949192431122706472553658.\end{aligned}$$

Then

$$H(P_5, Q_5, R_5, \rho_3^*) = -0.98\ldots, H(P_5, Q_5, R_5, \rho_4^*) = -1.02\ldots,$$
$$H(P_5, Q_5, R_5, \rho_5^*) = -2.10\ldots$$

with $\Delta_5 < 10^{-13}$. Therefore $H(P_5, Q_5, R_5, r) < 0$, $r \in [0,1]$.

(f) Using the *Mathematica* package "CountRoots", based on Sturm's method, we see that $G(P_6, Q_6, R_6, r)$ does not have zeros on $[0,1]$. Therefore $H(P_6, Q_6, R_6, r) < 0$, $r \in [0,1]$. ∎

We also need to discuss some equations of the form $H(P, Q, R, r) = 0$.

LEMMA A.2.2. *The following statements hold:*

(a) *If*

$$P_7(r) := -5 + 12r^2 - 30r^4 + 12r^6 - 5r^8,$$
$$Q_7(r) := 5r - 6r^3 + 5r^5, \quad R_7(r) = 5 - 6r^2 + 5r^4,$$

then the equation $H(P_7, Q_7, R_7, r) = 0$ *has the following two solutions on* $[0, 1]$

$$r_1 = 0.490314\ldots, \qquad r_2 = 0.549310\ldots,$$

and

(A.2.3) $$H(P_7, Q_7, R_7, r) = \begin{cases} < 0, & r \in [0, r_1) \cup (r_2, 1] \\ > 0, & r \in (r_1, r_2). \end{cases}$$

(b) *If*

$$P_8(r) := -625 + 1400r^2 - 7260r^4 + 13896r^6 - 19174r^8 + 13896r^{10}$$
$$- 7260r^{12} + 1400r^{14} - 625r^{16},$$
$$Q_8(r) := 500r - 1160r^3 + 4172r^5 - 4720r^7 + 4172r^9$$
$$- 1160r^{11} + 500r^{13},$$
$$R_8(r) := 5 - 6r^2 + 5r^4,$$

then the equation $H(P_8, Q_8, R_8, r) = 0$ *has the unique solution*

$$r_3 = 0.839169\ldots$$

on $[0, 1]$.

PROOF. We shall use again the *Mathematica* command N[Solve ...] to solve certain polynomial equations.

(a) It is easy to see that

(A.2.4) $$\begin{aligned}G(P_7, Q_7, R_7, r) &= (-5 + 22r^2 - 5r^4)(5 - 27r^2 + 55r^4 - 82r^6 \\ &\quad + 55r^8 - 27r^{10} + 5r^{12}).\end{aligned}$$

Each of polynomial factors in the right-hand side of (A.2.4) has the unique zero on $[0, 1]$. Solving the corresponding equations, we find the approximate values of r_1 and r_2.

Moreover, since

$$P_7(r) = -(5 - 2r^2 + r^4)(1 - 2r^2 + 5r^4) < 0, \quad Q_7(r) > 0, \qquad r \in [0, 1]$$

we have that r_1 and r_2 are the only zeros of $H(P_7, Q_7, R_7, r)$ on $[0, 1]$. Therefore (A.2.3) holds.

(b) By a straightforward calculation,

$$\begin{aligned}G(P_8, Q_8, R_8, r) &= (5 - 6r^2 + 5r^4)^3 (3125 - 12750r^2 + 77905r^4 \\ &\quad - 186280r^6 + 434298r^8 - 668436r^{10} + 434298r^{12} \\ &\quad - 186280r^{14} + 77905r^{16} - 12750r^{18} + 3125r^{20}).\end{aligned}$$

This polynomial has the unique zero r_3 on $[0, 1]$.

Next, for $r \in [0, 1]$

$$\begin{aligned}
P_8(r) &= (-5 + 16r - 26r^2 + 16r^3 - 5r^4) \\
&\quad \times (5 + 16r + 26r^2 + 16r^3 + 5r^4)(5 - 6r^2 + 5r^4)^2 < 0, \\
Q_8(r) &= 4r(5 - 6r^2 + 5r^4)(25 - 28r^2 + 150r^4 - 28r^6 + 25r^8) > 0.
\end{aligned}$$

Therefore r_3 is the unique zero of $H(P_8, Q_8, R_8, r)$ on $[0, 1]$. ∎

Next, we discuss some general properties of F_i. Recall that F_i is defined by (A.1.2).

LEMMA A.2.3. *For $i = 1, 2$, the following properties of F_i hold:*

(a) $\lim_{r \to 1-} F_i(r, t) = F_i(1, t) = 0, \qquad 0 \le t < \pi/2,$
(b) $\lim_{r \to 0+} F_i(r, t) = \infty, \qquad 0 \le t < \pi/2,$
(c) *For $r \in (0, 1)$ and $t \in [0, \pi/2)$,*

(A.2.5) $$\frac{\partial F_i(r, t)}{\partial r} = \frac{B_{1i}(r) \cos^2 2t + B_{2i}(r) \cos 2t + B_{3i}(r)}{r \cos t(1 + r^4 - 2r^2 \cos(2(t - \varphi_i)))(1 + r^4 - 2r^2 \cos(2(t + \varphi_i)))},$$

where

(A.2.6) $$\begin{aligned} B_{1i}(r) &:= 4r^4(1 + rf'_i(r)) - 2(1 + r^4)r^2 \cos(2\varphi_i(r)) \\ &\quad - 2r^3(1 - r^4)\varphi'_i(r) \sin(2\varphi_i(r)), \end{aligned}$$

(A.2.7) $$\begin{aligned} B_{2i}(r) &:= 4r^4 \cos^2(2\varphi_i(r)) + (1 + r^4)^2 \\ &\quad - 4(1 + r^4)r^2(1 - rf'_i(r)) \cos(2\varphi_i(r)), \end{aligned}$$

(A.2.8) $$\begin{aligned} B_{3i}(r) &:= 2r^3(1 - r^4)\varphi'_i(r) \sin(2\varphi_i(r)) - 2r^2(1 + r^4) \cos(2\varphi_i(r)) \\ &\quad + (1 + r^4)^2(1 + rf'_i(r)) \\ &\quad - 4r^2 \sin^2(2\varphi_i(r))(1 + rf'_i(r)). \end{aligned}$$

PROOF. Statements (a) and (b) immediately follow from (A.1.2)–(A.1.5). Next, by elementary manipulations, we have

$$\frac{\partial F_i(r,t)}{\partial r} = (1-r^4)\varphi_i'(r)\sin t \left(\frac{1}{1+r^4 - 2r^2\cos(2(t-\varphi_i))} \right.$$

$$\left. - \frac{1}{1+r^4 - 2r^2\cos(2(t+\varphi_i))} \right) - 2r \left(\frac{\sin(2(t-\varphi_i))}{1+r^4 - 2r^2\cos(2(t-\varphi_i))} \right.$$

$$\left. + \frac{\sin(2(t+\varphi_i))}{1+r^4 - 2r^2\cos(2(t+\varphi_i))} \right) + \frac{2\cos^2 t + r f_i'(r)}{r\cos t}$$

$$= \frac{\sin t \sin(2t)(4r^2(1-r^4)\varphi_i'(r)\sin(2\varphi_i) - 4r(1+r^4)\cos(2\varphi_i) + 8r^3\cos(2t))}{(1+r^4 - 2r^2\cos(2(t-\varphi_i)))(1+r^4 - 2r^2\cos(2(t+\varphi_i)))}$$

$$+ \frac{1+\cos(2t) + r f_i'(r)}{r\cos t}$$

$$= \left[r\cos t (1+r^4 - 2r^2\cos(2(t-\varphi_i)))(1+r^4 - 2r^2\cos(2(t+\varphi_i))) \right]^{-1}$$

$$\times \left[(1-\cos^2(2t))(2r^3(1-r^4)\varphi_i'\sin(2\varphi_i) - 2r^2(1+r^4)\cos(2\varphi_i) \right.$$

$$+ 4r^4\cos(2t)) + (1+\cos(2t) + r f_i'(r))((1+r^4)^2$$

$$\left. - 4(1+r^4)r^2\cos(2\varphi_i)\cos(2t) + 4r^4\cos^2(2t) - 4r^4\sin^2(2\varphi_i)) \right].$$

Hence (A.2.5) follows. ∎

Finally, we obtain a lower estimate for $F_2(r,t)$.

LEMMA A.2.4. *Let $[a,b] \subset (0,1)$ and $[c,d] \subseteq [0,\pi/2]$. Then*

$$\min_{\substack{r\in[a,b]\\t\in[c,d]}} F_2(r,t) \geq \sin d \big(A_1(a,b,d) + A_2(a,b,d) \big)$$

(A.2.9)
$$+ 2\cos c \log a + \left(\frac{1}{b} + b \right) \frac{\cos(\varphi_2(b))}{\cos c},$$

where

(A.2.10) $A_1(a,b,d) := \begin{cases} \arctan\left(\frac{1-b^2}{1+b^2} \cot(d+\varphi_2(a)) \right), & 0 \leq d + \varphi_2(a) \leq \frac{\pi}{2} \\ \arctan\left(\frac{1-a^2}{1+a^2} \cot(d+\varphi_2(a)) \right), & \pi/2 < d + \varphi_2(a) < \pi, \end{cases}$

(A.2.11) $A_2(a,b,d) := \begin{cases} \arctan\left(\frac{1-a^2}{1+a^2} \cot(d-\varphi_2(a)) \right), & 0 \leq d \leq \varphi_2(a) \\ \arctan\left(\frac{1-b^2}{1+b^2} \cot(d-\varphi_2(a)) \right), & 0 \leq \varphi_2(a) < d \leq 1. \end{cases}$

PROOF. We shall estimate each term separately in the right-side of (A.1.2) for $i = 2$.

We first note that the following elementary properties hold:
(i) For $r \in [0,1]$ and $\varphi \in [0,\pi/2]$, $K(r,t,\varphi)$ is decreasing in $t \in [0,\pi/2]$.
(ii) For $r \in [0,1]$, $t \in [0,\pi/2]$, and $\varphi \in [0,\pi/2]$, $K(r,t,\varphi) \leq 0$.
(iii) For $r \in [0,1)$ and $\varphi \in [0,\pi/2]$, $\sin t\, K(r,t,\varphi)$ is decreasing in $t \in [0,\pi/2]$.
(iv) For $r \in [0,1)$ and $t \in [0,\pi/2]$, $K(r,t,\varphi)$ is increasing in $\varphi \in [0,\pi/2]$.

(v) $\varphi_2(s) = \arctan \frac{1+s^2}{2(1-s^2)}$ is increasing in $s \in [0,1]$.

(vi) For $r \in [0,1]$ and $t \in [0,\pi/2]$, $K(r,t,\varphi_2(s))$ is increasing in $s \in [0,1]$.

(vii) For $t \in [0,\pi/2)$, $2\cos t \log r$ is increasing in $r \in (0,1]$, and for $r \in (0,1]$, it is increasing in $t \in [0,\pi/2]$.

(viii) For $t \in [0,\pi/2)$, $f_2(r)/\cos t$ is decreasing in $r \in (0,1]$, and for $r \in (0,1]$, it is increasing in $t \in [0,\pi/2)$.

(ix) $\arctan\bigl(A\frac{1-r^2}{1+r^2}\bigr)$ is decreasing in $r \in [0,1]$ if $A > 0$ and increasing in $r \in [0,1]$ if $A < 0$.

Properties (i), (v), (vii) and (ix) are obvious; (ii) follows from (i) since $K(r,0,\varphi) = 0$; and (iii) is an immediate consequence of (i) and (ii). Next, a straightforward calculation shows that

$$\frac{\partial K(r,t,\varphi)}{\partial \varphi} = \frac{4r^2(1-r^2)\sin 2t \sin 2\varphi}{(1+r^4 - 2r^2\cos(2(t-\varphi)))(1+r^4 - 2r^2\cos(2(t+\varphi)))}.$$

Hence property (iv) holds. Then (vi) follows from (iv) and (v). Finally, (v) implies (viii).

To estimate $\sin t\, K(r,t,\varphi_2(r))$, we note that by (iii), (vi), and (ix),

$$\min_{\substack{r \in [a,b] \\ t \in [c,d]}} \sin t\, K(r,t,\varphi_2(r)) \geq \min_{r \in [a,b]} \min_{\substack{s \in [a,b] \\ t \in [c,d]}} \sin t\, K(r,t,\varphi_2(s))$$

$$\geq \min_{r \in [a,b]} \sin d\, K(r,d,\varphi_2(a))$$

(A.2.12) $$\geq \sin d\bigl(A_1(a,b,d) + A_2(a,b,d)\bigr).$$

Further by (vii) and (viii),

$$\min_{\substack{r \in [a,b] \\ t \in [c,d]}} 2\cos t \log r + \min_{\substack{r \in [a,b] \\ t \in [c,d]}} (1/r + r)\cos(\varphi_2(r))$$

(A.2.13) $$\geq 2\cos c \log a + \left(\frac{1}{b} + b\right)\frac{\cos(\varphi_2(b))}{\cos c}.$$

Thus (A.2.12) and (A.2.13) yield (A.2.9). ∎

A.3. Proof of Proposition A.1.1

<u>Case $i = 1$.</u> In this case, we have

$$\varphi_1(r) = \arctan \frac{2(1+r^2)}{1-r^2}, \qquad \varphi_1'(r) = \frac{8r}{5 + 6r^2 + 5r^4},$$

$$\sin(\varphi_1(r)) = \frac{2(1+r^2)}{\sqrt{5+6r^2+5r^4}}, \qquad \cos(\varphi_1(r)) = \frac{1-r^2}{\sqrt{5+6r^2+5r^4}},$$

$$\sin(2\varphi_1(r)) = \frac{4(1-r^4)}{5+6r^2+5r^4}, \qquad \cos(2\varphi_1(r)) = -\frac{3+10r^2+3r^4}{5+6r^2+5r^4},$$

$$f_1(r) = \frac{2(1-r^4)}{r\sqrt{5+6r^2+5r^4}},$$

$$f_1'(r) = -\frac{2(5+12r^2+30r^4+12r^6+5r^8)}{r^2(5+6r^2+5r^4)^{3/2}}.$$

Using these formulae and relations (A.2.5)–(A.2.8), we obtain

(A.3.1)
$$\frac{\partial F_1(r,t)}{\partial r} = \frac{B_{11}(r)\cos^2(2t) + B_{21}(r)\cos(2t) + B_{31}(r)}{r\cos t(1 + r^4 - 2r^2\cos(2(t-\varphi_1)))(1 + r^4 - 2r^2\cos(2(t+\varphi_1)))},$$

where

$$B_{11}(r) = \frac{2r^2(5 + 2r^2 + r^4)(1 + 2r^2 + 5r^4)(3 + 10r^2 + 3r^4 - 4r\sqrt{5 + 6r^2 + 5r^4})}{(5 + 6r^2 + 5r^4)^2},$$

$$\begin{aligned}
B_{21}(r) &= (5 + 2r^2 + r^4)(1 + 2r^2 + 5r^4)\bigl(-24r - 80r^3 - 48r^5 \\
&\quad - 80r^7 - 24r^9 + (5 + 12r^2 + 30r^4 + 12r^6 + 5r^8)\sqrt{5 + 6r^2 + 5r^4}\bigr) \\
&\quad \times (5 + 6r^2 + 5r^4)^{-5/2},
\end{aligned}$$

$$\begin{aligned}
B_{31}(r) &= (5 + 2r^2 + r^4)(1 + 2r^2 + 5r^4)\bigl(-50 - 120r^2 - 144r^4 - 360r^6 \\
&\quad - 700r^8 - 360r^{10} - 144r^{12} - 120r^{14} - 50r^{16} \\
&\quad + \bigl(25r + 60r^3 + 111r^5 + 120r^7 + 111r^9 + 60r^{11} + 25r^{13}\bigr) \\
&\quad \times \sqrt{5 + 6r^2 + 5r^4}\bigr)r^{-1}(5 + 6r^2 + 5r^4)^{-7/2}.
\end{aligned}$$

We establish the inequality $F_1(r,t) \geq 0$ by proving that $\frac{\partial F_1(r,t)}{\partial r} < 0$ for $r \in (0,1)$ and $t \in [0, \pi/2)$, i.e., $F_1(r,t)$ is decreasing in $r \in (0,1)$. To show this, it suffices to prove the inequality

(A.3.2) $\quad H_1(r,y) := B_{11}(r)y^2 + B_{21}(r)y + B_{31}(r) < 0, \quad r \in (0,1), \ y \in [-1,1]$

(see (A.3.1)).

Note first that by Lemma A.2.1(a),

(A.3.3) $\qquad\qquad B_{11}(r) > 0, \qquad r \in (0,1),$

since $\text{sign}(B_{11}(r)) = \text{sign}(P_1(r) + Q_1(r)\sqrt{R_1(r)})$, $r \in (0,1)$. Further, the following inequalities are valid:

(A.3.4) $\qquad\qquad H_1(r, \pm 1) < 0, \qquad r \in (0,1).$

Indeed, a straight computation shows that

$$\begin{aligned}
H_1(r,-1) &= B_{11}(r) - B_{21}(r) + B_{31}(r) \\
&= \frac{-50(1-r^4)^4(5 + 2r^2 + r^4)(1 + 2r^2 + 5r^4)}{r(5 + 6r^2 + 5r^4)^{7/2}}.
\end{aligned}$$

Hence $H_1(r,-1) < 0$ for $r \in (0,1)$. Next by elementary manipulations,

$$\begin{aligned}
H_1(r,1) &= B_{11}(r) + B_{21}(r) + B_{31}(r) = 2(5 + 2r^2 + r^4)^2(1 + 2r^2 + 5r^4)^2 \\
&\quad \times \bigl(-5 - 12r^2 - 30r^4 - 12r^6 - 5r^8 \\
&\quad + (5r + 6r^3 + 5r^5)\sqrt{5 + 6r^2 + 5r^4}\bigr) \times r^{-1}(5 + 6r^2 + 5r^4)^{-7/2}.
\end{aligned}$$

Then $\text{sign}(H_1(r,1)) = \text{sign}(P_2(r) + Q_2(r)\sqrt{R_2(r)})$, and by Lemma A.2.1(b), we have $H_1(r,1) < 0$ for $r \in (0,1)$. Thus (A.3.4) holds.

To complete the proof of inequality (A.1.6) for $i = 1$, it suffices to use statements (a) and (b) of Lemma A.2.3.

Case $i = 2$. In this case, we have
$$\varphi_2(r) = \arctan\frac{1+r^2}{2(1-r^2)}, \quad \varphi_2'(r) = \frac{8r}{5 - 6r^2 + 5r^4},$$

$$\sin(\varphi_2(r)) = \frac{1+r^2}{\sqrt{5 - 6r^2 + 5r^4}}, \quad \cos(\varphi_2(r)) = \frac{2(1-r^2)}{\sqrt{5 - 6r^2 + 5r^4}},$$

$$\sin(2\varphi_2(r)) = \frac{4(1-r^4)}{5 - 6r^2 + 5r^4}, \quad \cos(2\varphi_2(r)) = -\frac{3 - 10r^2 + 3r^4}{5 - 6r^2 + 5r^4},$$

$$f_2(r) = \frac{2(1-r^4)}{r\sqrt{5 - 6r^2 + 5r^4}},$$

$$f_2'(r) = -\frac{2(5 - 12r^2 + 30r^4 - 12r^6 + 5r^8)}{r^2(5 - 6r^2 + 5r^4)^{3/2}}.$$

Using these formulae and relations (A.2.5)–(A.2.8), we obtain

(A.3.5) $\quad \dfrac{\partial F_2(r,t)}{\partial r} =$
$$\frac{B_{12}\cos^2(2t) + B_{22}(r)\cos 2t + B_{32}(r)}{r\cos t\bigl(1 + r^4 - 2r^2\cos(2(t-\varphi_2))\bigr)\bigl(1 + r^4 - 2r^2\cos(2(t+\varphi_2))\bigr)},$$

where

$$\begin{aligned}
B_{12}(r) &= 2r^2(5 - 2r^2 + r^4)(1 - 2r^2 + 5r^4)\bigl(-20r + 24r^3 - 20r^5 \\
&\quad + (-3 + 10r^2 - 3r^4)\sqrt{5 - 6r^2 + 5r^4}\bigr)(5 - 6r^2 + 5r^4)^{-5/2}, \\
B_{22}(r) &= (5 - 2r^2 + r^4)(1 - 2r^2 + 5r^4)\bigl(24r - 80r^3 + 48r^5 - 80r^7 + 24r^9 \\
&\quad + (5 - 12r^2 + 30r^4 - 12r^6 + 5r^8) \\
&\quad \times \sqrt{5 - 6r^2 + 5r^4}\bigr)(5 - 6r^2 + 5r^4)^{-5/2}, \\
B_{32}(r) &= (5 - 2r^2 + r^4)(1 - 2r^2 + 5r^4)\bigl(-50 + 120r^2 - 144r^4 + 360r^6 - 700r^8 \\
&\quad + 360r^{10} - 144r^{12} + 120r^{14} - 50r^{16} + (25r - 60r^3 + 111r^5 - 120r^7 \\
&\quad + 111r^9 - 60r^{11} + 25r^{13})\sqrt{5 - 6r^2 + 5r^4}\bigr)r^{-1}(5 - 6r^2 + 5r^4)^{-7/2}.
\end{aligned}$$

We first discuss the behavior of the function
$$H_2(r,y) := B_{12}(r)y^2 + B_{22}(r)y + B_{32}(r), \qquad r \in (0,1).$$

Unlike $H_1(r,y)$, the function $H_2(r,y)$ is not negative for all relevant r and y, i.e., the method of the proof of Proposition A.1.1 for $i = 1$ is not applicable to the case of $i = 2$. We can also use a computational procedure based on Lemma A.2.4 to prove the inequality $F_2(r,t) \geq 0$, but this method fails in some neighborhoods of the sets $\{(r,t):\ r = 0,\ 0 \leq t < \pi/2\}$ and $\{(r,t):\ r = 1,\ 0 \leq t \leq \pi/2\}$, since F_2 is too large or too small in these neighborhoods. That is why we divide $[0,1]$ into three intervals $[0,\rho_1]$, $[\rho_1,\rho_3]$ and $[\rho_3,1]$ and prove the inequality $F_2(r,t) \geq 0$ by the computational procedure for $r \in [\rho_1,\rho_3]$. Proofs of this inequality for other two intervals are based on properties of $H_2(r,t)$ given below.

Let us recall that the numbers
$$\rho_1 = 0.490314\ldots, \quad \rho_2 = 0.549310\ldots, \quad \rho_3 = 0.839160\ldots$$
are introduced in Lemma A.2.2.

LEMMA A.3.1. *The following properties of $H_2(r,y)$ hold:*
(a) *For $r \in [0,1]$, $B_{12}(r) < 0$.*
(b) *For each $r \in (0,1]$, the equation*

(A.3.6) $$H_2(r,y) = 0$$

has two real solutions

(A.3.7) $$y_1(r) := \frac{-B_{22}(r) + \sqrt{B_{22}^2(r) - 4B_{12}(r)B_{32}(r)}}{2B_{12}(r)},$$

(A.3.8) $$y_2(r) = \frac{-B_{22}(r) - \sqrt{B_{22}^2(r) - 4B_{12}(r)B_{32}(r)}}{2B_{12}(r)}$$

with the properties:
(b1) $y_1(r) < y_2(r)$, $r \in (0,1)$;
(b2) $y_i(1) = -1$, $\lim_{r \to 0+} y_i(r) = \infty$, $i = 1, 2$;
(b3) y_i *is decreasing in $r \in (0,1]$, $i = 1, 2$.*
(c) *For $r \in (0,1)$, $H_2(r, -1) < 0$.*
(d)

(A.3.9) $$H_2(r, 1) \begin{cases} < 0, & r \in (0, \rho_1) \\ > 0, & r \in (\rho_1, \rho_2) \\ < 0, & r \in (\rho_2, 1) \\ = 0, & r \in \{\rho_1, \rho_2\}. \end{cases}$$

(e) *For $i = 1, 2$,*

(A.3.10) $$y_i(r) \begin{cases} \in (1, \infty), & r \in (0, \rho_i) \\ = 1, & r = \rho_i \\ \in [-1, 1], & \rho \in (\rho_i, 1]. \end{cases}$$

(f) *For $T_i(r) := (1/2)\arccos(y_i(r))$, $i = 1, 2$, we have*

(A.3.11) $$\varphi_2(r) - T_1(r) \begin{cases} > 0, & r \in [\rho_1, \rho_3) \\ < 0, & r \in (\rho_3, 1) \\ = 0, & r \in \{\rho_3, 1\}. \end{cases}$$

(g) *For $r \in [\rho_3, 1)$,*

(A.3.12) $$0 < T_1(r) - \varphi_2(r) < \arctan\left(\frac{0.07(1-r^2)}{1+r^2}\right).$$

PROOF. (a) It is easy to see that for $r \in (0,1]$,

$$\operatorname{sign} B_{12}(r) = \operatorname{sign}\bigl(P_3(r) + Q_3(r)\sqrt{R_3(r)}\bigr).$$

Hence by Lemma A.2.1(c), $B_{12}(r) < 0$, $r \in (0,1]$.

(b) By elementary manipulations,

$$\begin{aligned}
B(r) &:= B_{22}^2(r) - 4B_{12}(r)B_{32}(r) = (1-r^4)^4(5 - 2r^2 + r^4)^2(1 - 2r + 5r^4)^2 \\
&\quad \times \bigl(2560r^3 + (125 - 1174r^2 + 125r^4) \\
&\quad \times \sqrt{5 - 6r^2 + 5r^4}\bigr)(5 - 6r^2 + 5r^4)^{-5/2}.
\end{aligned}$$

Since for $r \in [0,1)$,
$$\operatorname{sign} B(r) = \operatorname{sign}\bigl(P_4(r) + Q_4(r)\sqrt{R_4(r)}\bigr),$$
we have from Lemma A.2.1(d) that $B(r) > 0$, $r \in (0,1)$, and $B(1) = 0$. Therefore, for each $r \in (0,1]$, equation (A.3.6) has two real solutions $y_1(r)$ and $y_2(r)$ given in (A.3.7) and (A.3.8).

It remains to prove the properties of $y_i(r)$, $i = 1, 2$. Property (b1) immediately follows from statement (a), while (b2) is an easy consequence of (A.3.7) and (A.3.8).

To prove (b3), we first note that for $r \in (0, 1)$,

(A.3.13)
$$\frac{dy_i(r)}{dr} = -\frac{B'_{12}(r)y_i^2(r) + B'_{22}(r)y_i(r) + B'_{32}(r)}{2y_i(r)B_{12}(r) + B_{22}(r)},$$

where by statements (a) and (b1),

(A.3.14)
$$\operatorname{sign}\bigl(2y_i(r)B_{12}(r) + B_{22}(r)\bigr) = (-1)^{i+1}, \qquad i = 1, 2.$$

Next, by a straightforward calculation, we have

$$\begin{aligned}
B'_{12}(r) &= 4r\bigl(-750r + 3900r^3 - 15540r^5 + 28428r^7 \\
&\quad - 34160r^9 + 26484r^{11} - 14860r^{13} + 5700r^{15} - 1250r^{17} \\
&\quad + \bigl(-75 + 770r^2 - 3150r^4 + 7930r^6 - 8964r^8 + 6630r^{10} \\
&\quad - 3714r^{12} + 1310r^{14} - 225r^{16}\bigr)
\end{aligned}$$

(A.3.15)
$$\times \sqrt{5 - 6r^2 + 5r^4}\bigr)(5 - 6r^2 + 5r^4),$$

$$\begin{aligned}
B'_{22}(r) &= 8\bigl(75 - 930r^2 + 4293r^4 - 13020r^6 + 19746r^8 \\
&\quad - 29120r^{10} + 29838r^{12} - 20260r^{14} + 10035r^{16} - 3230r^{18} \\
&\quad + 525r^{20} + \bigl(-75r + 805r^3 - 27000r^5 + 6340r^7 - 7914r^9 + 7110r^{11} \\
&\quad - 4764r^{13} + 2260r^{15} - 675r^{17} + 125r^{19}\bigr)\sqrt{5 - 6r^2 + 5r^4}\bigr)
\end{aligned}$$

(A.3.16)
$$\times (5 - 6r^2 + 5r^4)^{-7/2},$$

$$\begin{aligned}
B'_{32}(r) &= 2\bigl(625 - 3000r^2 + 3525r^4 + 13680r^6 - 95007r^8 \\
&\quad + 351000r^{10} - 752643r^{12} + 839520r^{14} - 705429r^{16} \\
&\quad + 602040r^{18} - 440265r^{20} + 252720r^{22} - 110525r^{24} \\
&\quad + 33000r^{26} - 5625r^{28} + \bigl(-750r^3 + 9300r^5 - 33330r^7 \\
&\quad + 82900r^9 - 148140r^{11} + 205768r^{13} - 217860r^{15} \\
&\quad + 177768r^{17} - 107910r^{19} + 48100r^{21} - 14250r^{23}
\end{aligned}$$

(A.3.17)
$$+ 2500r^{25}\bigr)\sqrt{5 - 6r^2 + 5r^4}\bigr)r^{-2}(5 - 6r^2 + 5r^4)^{-9/2}.$$

Then using (A.3.7), (A.3.8) and (A.3.15)–(A.3.17), we obtain

$$\begin{aligned}
&16B_{12}^2\bigl(B'_{12}(r)y_1^2(r) + B'_{22}y_1(r) + B'_{32}(r)\bigr)\bigl(B'_{12}(r)y_2^2(r) \\
&\quad + B'_{22}(r)y_2(r) + B'_{32}(r)\bigr) \\
&= \bigl(2B'_{12}(r)B_{22}^2(r) - 4B'_{12}(r)B_{12}(r)B_{32}(r) - 2B'_{22}(r)B_{12}(r)B_{22}(r) \\
&\quad + 4B'_{32}(r)B_{12}^2(r)\bigr)^2 - \bigl(2B'_{22}(r)B_{12}(r) - 2B'_{12}(r)B_{22}(r)\bigr)^2 \\
&\quad \times \bigl(B_{22}^2(r) - 4B_{12}(r)B_{32}(r)\bigr) = 12800(1 - r^4)^6(5 - 2r^2 + r^4)^6
\end{aligned}$$

(A.3.18)
$$\times (1 - 2r^2 + 5r^4)^6\bigl(P_5(r) + Q_5(r)\sqrt{R_5(r)}\bigr).$$

Since by Lemma A.2.1(e), $P_5(r) + Q_5(r)\sqrt{R_5(r)} < 0$ for all $r \in [0,1]$, we have from (A.3.13), (A.3.14), and (A.3.18) that $y_1'(r)y_2'(r) > 0$, $r \in [0,1]$. This implies that either $y_i'(r) < 0$, $i = 1, 2$, for all $r \in (0, 1)$, or $y_i'(r) > 0$, $i = 1, 2$, for all $r \in (0, 1)$. The latter statement contradicts to property (b2) so (b3) holds.

(c) By a straightforward calculation, we have
$$H_2(r,-1) = -50r^{-1}(5 - 6r^2 + 5r^4)^{-7/2}(1-r^4)^4(1-2r^2+5r^4)$$
$$\times (5 - 2r^2 + r^4) < 0, \quad r \in (0,1).$$

(d) By a straightforward calculation,
$$H_2(r,1) = 2r^{-1}(5 - 6r^2 + 5r^4)^{-7/2}(5 - 2r^2 + r^4)^2(1 - 2r^2 + 5r^4)^2$$
$$\times \left(-5 + 12r^2 - 30r^4 + 12r^6 - 5r^8 \right.$$
$$\left. + (5r - 6r^3 + 5r^5)\sqrt{5 - 6r^2 + 5r^4}\right).$$

Since
$$\operatorname{sign} H_2(r,1) = \operatorname{sign}\left(P_7(r) + Q_7(r)\sqrt{R_7(r)}\right),$$
we see that statement (d) follows from Lemma A.2.2(a).

(e) This property follows from (b) and (d).

(f) Property (e) shows that $T_i(r)$ is defined on $[\rho_i, 1]$. Note that

(A.3.19) $$\varphi_2(\rho_1) - T_1(\rho_1) = 0.68\ldots.$$

Next, we find all solutions to the equations

(A.3.20) $$\varphi_2(r) - T_i(r) = 0, \quad r \in [\rho_i, 1), \quad i = 1, 2.$$

These equations are equivalent to

(A.3.21) $$y_i(r) = \frac{3 - 10r^2 + 3r^4}{5 - 6r^2 + 5r^4}, \quad r \in [\rho_i, 1), \quad i = 1, 2.$$

Substituting y in (A.3.6) with the right-hand side of (A.3.21), we conclude that every solution to equations (A.3.20) satisfies the equation

(A.3.22) $$H_2\left(r, \frac{3 - 10r^2 + 3r^4}{5 - 6r^2 + 5r^4}\right) = 0, \quad r \in (0,1).$$

Further, a straightforward calculation shows that
$$H_2\left(r, \frac{3 - 10r^2 + 3r^4}{5 - 6r^2 + 5r^4}\right) = 2(1-r^2)^4(5 - 2r^2 + r^4)(1 - 2r^2 + 5r^4)$$
$$\times \left(r^{-1}(5 - 6r^2 + 5r^4)^{-11/2}\right)\left(P_8(r) + Q_8(r)\sqrt{R_8(r)}\right).$$

Using Lemma A.2.2(b), we obtain that equation (A.3.22) has the unique solution $r = \rho_3$ on $(0, 1)$. Since $\rho_3 > \rho_2 > \rho_1$, we have that the union of the solution sets to two equations (A.3.20) is $\{\rho_3\}$.

Furthermore, we notice that if $\varphi_2(\rho_3) - T_2(\rho_3) = 0$, then $\varphi_2(\rho_3) - T_1(\rho_3) < 0$, by (b1). Combining this inequality with (A.3.19), we conclude that $\varphi_2 - T_1 = 0$ has a solution on (ρ_1, ρ_3). This contradiction shows that $\varphi_2 - T_2$ has no zeros on $(\rho_2, 1)$, while the only zero of $\varphi_2 - T_1$ on $(\rho_1, 1)$ is $r = \rho_3$.

Finally, (A.3.19) yields (A.3.11).

(g) The left inequality in (A.3.12) follows from (A.3.11). To prove the right one, we consider the equation

$$(A.3.23) \qquad M_c(r) := T_1(r) - \varphi_2(r) - \arctan\left(\frac{2c(1-r^2)}{1+r^2}\right) = 0, \quad r \in [\rho_3, 1),$$

where c is a fixed number from $(0,1)$. Then it is easy to verify that

$$\begin{aligned} M_c(r) &= \arctan\left(\sqrt{\frac{1-y_1(r)}{1+y_1(r)}}\right) - \arctan\left(\frac{1+r^2}{2(1-r^2)}\right) - \arctan\left(\frac{2c(1-r^2)}{1+r^2}\right) \\ &= \arctan\left(\sqrt{\frac{1-y_1(r)}{1+y_1(r)}}\right) - \arctan\left(\frac{1+4c+2(1-4c)r^2+(1+4c)r^4}{2(1-c)(1-r^4)}\right). \end{aligned}$$

Hence choosing $c = 35/1000$, we have that for $r \in (\rho_3, 1)$, equation (A.3.23) is equivalent to $y_1(r) = g(r)$, where

$$g(r) := \frac{24253 - 39216r^2 - 117078r^4 - 39216r^6 + 24253r^8}{50245 + 39216r^2 - 31918r^4 + 39216r^6 + 50245r^8}.$$

Substituting y in (A.3.6) with $g(r)$, we conclude that every solution to (A.3.23) with $c = 35/1000$ satisfies the equation

$$(A.3.24) \qquad H_2(r, g(r)) = 0.$$

Next,

$$\begin{aligned} H_2(r, g(r)) &= 2(1-r^4)^4(5 - 2r^2 + r^4)(1 - 2r^2 + 5r^4)r^{-1}(5 - 6r^2 + 5r^4)^{-7/2} \\ &\quad \times \left(P_6(r) + Q_6(r)\sqrt{R_6(r)}\right). \end{aligned}$$

Then by Lemma A.2.1(f), equation (A.3.24) has no solutions in $(0,1)$. Therefore $M_{0.035}(r)$ has no zeros in $[\rho_3, 1)$. Since $M_{0.035}(\rho_3) < 0$, statement (g) is established. This completes the proof of Lemma A.3.1. ∎

We now are in position to prove Proposition A.1.1 for $i = 2$. The proof consists of three steps, corresponding to different subintervals of $[0, 1]$.

$r \in [0.49, 0.84]$. Let us set $A = 0.49$, $B = 0.84$,

$$a_n := A + \frac{(B-A)(n-1)}{N}, \quad c_m := \frac{\pi(m-1)}{2M}, \quad 1 \le n \le N+1, \ 1 \le m \le M+1,$$

where N and M are natural numbers. Then

$$[0.49, 0.84] = \bigcup_{n=1}^{N} [a_n, a_{n+1}], \qquad [0, \pi/2] = \bigcup_{m=1}^{M} [c_m, c_{m+1}],$$

and by Lemma A.2.4

$$\begin{aligned} \min_{\substack{r \in [0.49, 0.84] \\ t \in [0, \pi/2]}} F_2(r, t) &= \min_{\substack{1 \le n \le N \\ 1 \le m \le M}} \min_{\substack{r \in [a_n, a_{n+1}] \\ t \in [c_m, c_{m+1}]}} F_2(r, t) \\ &\ge \min_{\substack{1 \le n \le N \\ 1 \le m \le M}} \big(\sin c_{m+1}(A_1(a_n, a_{n+1}, c_{m+1}) + A_2(a_n, a_{n+1}, c_{m+1})) \\ &\quad + (1/a_{n+1} + a_{n+1})\cos(\varphi_2(a_{n+1}))/\cos(c_m)\big), \end{aligned}$$
(A.3.25)

where A_1 and A_2 are defined at (A.2.10) and (A.2.11).

Finally, choosing $N = 80$, $M = 125$, and computing the right-hand side of (A.3.25) by the *Mathematica* command Min, we obtain from (A.3.25) that

$$\text{(A.3.26)} \qquad \min_{\substack{r \in [0.49, 0.84] \\ t \in [0, \pi/2]}} F_2(r, t) \geq 0.0903 \ldots .$$

$r \in [0, \rho_1]$. It follows from statements (c) and (e) of Lemma A.3.1 that $H_2(r, y) \leq 0$ for all $r \in (0, \rho_1]$ and $y \in [-1, 1]$. Since by (A.3.5),

$$\text{sign}\left(\frac{\partial F_2(r,t)}{\partial r}\right) = \text{sign}(H_2(r, \cos(2t))),$$

we conclude that $F_2(r, t)$ is decreasing in $r \in (0, \rho_1]$ for each $t \in [0, \pi/2)$.

Then using Lemma A.2.3(b) and estimate (A.3.26), we have

$$\text{(A.3.27)} \qquad \inf_{\substack{r \in (0, \rho_1] \\ t \in [0, \pi/2)}} F_2(r, t) \geq \min_{t \in [0, \pi/2]} F_2(\rho_1, t) \geq 0.0903 \ldots .$$

$r \in [\rho_3, 1]$. Let us define $F_2(r, \pi/2) := 0$, $r \in [\rho_3, 1]$. Then F_2 is lower semi-continuous on the rectangle

$$\Pi := \{(r, t) \in \mathbb{R}^2 : r \in [\rho_3, 1], t \in [0, \pi/2]\},$$

i.e., $F(r, t) \leq \liminf_{n \to \infty} F_2(r_n, t_n)$ for any $(r, t) \in \Pi$ and every sequence $(r_n, t_n) \in \Pi$, $n = 1, 2, \ldots$, such that $\lim_{n \to \infty} (r_n, t_n) = (r, t)$. Indeed, $F_2(r, t)$ is continuous at any $(r, t) \in \Pi$ with $t \in [0, \pi/2)$ and semi-continuous at $(r, \pi/2)$, $r \in [\rho_3, 1]$, since by (A.1.3), $K(r, \pi/2, \varphi_2(r))$ is continuous on $[\rho_3, 1]$ and $f_2(r)/\cos t \geq 0$ on Π.

Then by the Weierstrass theorem [24, 109], there exists $(r_0, t_0) \in \Pi$ such that $\min_{(r,t) \in \Pi} F_2(r, t) = F_2(r_0, t_0)$.

We first assume that (r_0, t_0) belongs to the boundary $\partial \Pi$ of Π. Then by (A.3.26),

$$\text{(A.3.28)} \qquad \min_{t \in [0, \pi/2]} F_2(\rho_3, t) \geq 0.0903 \ldots .$$

Next, using Lemma A.3.1(d), we have that $\partial F_2(r, 0)/\partial r < 0$ for all $r \in [\rho_3, 1)$. This implies

$$\text{(A.3.29)} \qquad \min_{r \in [\rho_3, 1]} F_2(r, 0) \geq F_2(1, 0) = 0.$$

Since $F_2(1, t) = 0$, $t \in [0, \pi/2)$, and $f_2(r, \pi/2) = 0$, $r \in [\rho_3, 1]$, we see that (A.3.28) and (A.3.29) yield the inequality

$$\text{(A.3.30)} \qquad \min_{(r,t) \in \Pi} F_2(r, t) = \min_{(r,t) \in \partial \Pi} F_2(r, t) \geq 0.$$

Suppose now that (r_0, t_0) belongs to the interior of Π. Then $\partial F_2(r_0, t_0)/\partial r = 0$, so by Lemma A.3.1(b), $(r_0, t_0) \in \Gamma_1 \cup \Gamma_2$, where $\Gamma_i := \{(r, T_i(r)) \in \mathbb{R}^2 : \rho_3 < r < 1\}$, $i = 1, 2$, are continuous curves in Π. We recall that $T_i(r) = (1/2) \arccos(y_i(r))$, $i = 1, 2$, are defined in Lemma A.3.1. Since due to statement (b) of Lemma A.3.1, $T_1(r) > T_2(r)$, $\rho_3 < r < 1$, we can introduce the following sets:

$$S_1 := \{(r, t) \in \Pi : t > T_1(r)\}, \quad S_2 := \{(r, t) \in \Pi : T_2(r) < t < T_1(r)\},$$
$$S_3 := \{(r, t) \in \Pi : t < T_2(r)\}.$$

It follows from statements (c), (d), and (e) of Lemma A.3.1 that

(A.3.31) $$\frac{\partial F_2(r,t)}{\partial r} \begin{cases} < 0, & (r,t) \in S_1 \\ > 0, & (r,t) \in S_2 \\ < 0, & (r,t) \in S_3. \end{cases}$$

Using the fact that $T_i(r)$ is increasing in $r \in (\rho_3, 1)$ (see statement (b3) of Lemma A.3.1), we obtain from (A.3.31) that for every $(r,t) \in \Gamma_i$ there exists $\varepsilon > 0$ such that for all $u \in (0, \varepsilon)$,

$$\text{sign}\left(\frac{\partial F_2(r-u,t)}{\partial r}\right) = (-1)^i, \ \text{sign}\left(\frac{\partial F_2(r+u,t)}{\partial r}\right) = (-1)^{i+1}, \ i = 1, 2.$$

This shows that $(r_0, t_0) \notin \Gamma_2$. Thus it suffices to estimate F_2 on Γ_1. We recall that $F_2(r_0, t_0) = F_2(r_0, T_1(r_0))$. We also recall that by Lemma A.3.1(g),

(A.3.32) $$0 \leq T_1(r) - \varphi_2(r) < \arctan\left(\frac{2c(1-r^2)}{1+r^2}\right), \ r \in [\rho_3, 1),$$

where here and in the sequel, $c = 0.035$. Hence for $r \in [\rho_3, 1)$,

$$\begin{aligned} F_2(r, T_1(r)) &= \sin(T_1(r))\left[\arctan\left(\frac{1-r^2}{1+r^2}\cot(T_1(r) + \varphi_2(r))\right)\right. \\ &\quad \left. + \arctan\left(\frac{1-r^2}{1+r^2}\cot(T_1(r) - \varphi_2(r))\right) - \pi\right] + 2\cos(T_1(r))\log r \\ &\quad + \left(\frac{1}{r} + r\right)\frac{\cos(\varphi_2(r))}{\cos(T_1(r))} \\ &\geq \arctan\left(\frac{1-r^2}{1+r^2}\cot\left(2\varphi_2(r) + \arctan\left(\frac{2c(1-r^2)}{1+r^2}\right)\right)\right) \end{aligned}$$

(A.3.33) $\qquad + \arctan(1/(2c)) + 2\cos(\varphi_2(r))\log r + 1/r + r - \pi.$

Next by elementary manipulations,

$$2\varphi_2(r) + \arctan\left(\frac{2c(1-r^2)}{1+r^2}\right) = 2\arctan\left(\frac{1+r^2}{2(1-r^2)}\right) + \arctan\left(\frac{2c(1-r^2)}{1+r^2}\right)$$

(A.3.34) $\quad = \pi + \arctan\left(\frac{4(1-r^4)}{3 - 10r^2 + 3r^4}\right) + \arctan\left(\frac{2c(1-r^2)}{1+r^2}\right)$

$\quad = \pi + \arctan\left(\frac{(1-r^2)(4 + 6c + (8-20c)r^2 + (4+4c)r^4)}{(1+r^2)(3 - 8c + (-10+16c)r^2 + (3-8c)r^4)}\right).$

Further taking account of the inequalities

$$\cos(\varphi_2(r))\log r \geq \cos(\varphi_2(\rho_3))\log r, \quad 1/r + r \geq 1, \ r \in [\rho_3, 1],$$

and substituting c with $35/1000$ in (A.3.34), we have from (A.3.33) and (A.3.34)

(A.3.35) $\quad F_2(r, T_1(r)) \geq \arctan(G(r)) + 0.656146\log r + 0.359317, \quad r \in [\rho_3, 1),$

where

$$G(r) := \frac{16(17 - 59r^2 + 17r^4)}{421 + 730r^2 + 421r^4}.$$

Since

$$G'(r) = \frac{1191968(r^4 - 1)r}{(421 + 730r^2 + 421r^4)^2},$$

we see that $\arctan(G(r))$ is decreasing in $[\rho_3, 1]$. Thus (A.3.35) implies

$$\inf_{r \in [\rho_3, 0.86]} F_2(r, T_1(r)) \geq \arctan(G(0.86)) + 0.656146 \log 0.8391$$
$$+ 0.359317 > 0.0153,$$
$$\inf_{r \in [0.86, 1]} F_2(r, T_1(r)) \geq \arctan(G(1)) + 0.656146 \log 0.86$$
$$+ 0.359317 > 0.0111.$$

Combining these inequalities with (A.3.30), we obtain

(A.3.36) $$\inf_{\substack{r \in [\rho_3, 1] \\ t \in [0, \pi/2)}} F_2(r, t) \geq \min_{\substack{r \in [\rho_3, 1] \\ t \in [0, \pi/2]}} F_2(r, t) \geq 0.$$

Finally, (A.3.26), (A.3.27), and (A.3.36) yield (A.1.6) for $i = 2$. This completes the proof of Proposition A.1.1. ∎

Bibliography

[1.] Abramovitz, M. and Stegun, I. A., *Handbook of Mathematical Functions with Formulas, Graphs, and Mathematical Tables*, Dover, New York, 1972.

[2.] Ahundov, A., On best approximation in the mean of the functions $F_{s,m}(|a-x|) = |a-x|^s \ln^m |a-x|$, *Proc. Azerbaijan Ped. Inst.*, **2** (1955), 117–132. (In Russian)

[3.] Akhiezer, N. I., On weighted approximation of continuous functions by polynomials on the entire real axis, *AMS Transl.*, Ser. 2, **22** (1962), 95–137.

[4.] Akhiezer, N. I., *Lectures on the Theory of Approximation*, 2nd Edition, Nauka, Moscow, 1965. (In Russian)

[5.] Akhiezer, N. I. and Bernstein, S. N., Generalization of a theorem on weight functions and application to the problem of moments, *Dokl. Akad. Nauk SSSR*, **92** (1953), 1109–1112. (In Russian)

[6.] Akhiezer, N. I. and Krein, M. G., On the best approximation of periodic functions, *Dokl. Akad. Nauk SSSR*, **15** (1937), 107–112. (In Russian)

[7.] Bernstein, S., Sur les meilleure approximation de $|x|$ par des polynômes des degrés donnés, *Acta Math.*, **37** (1913), 1–57.

[8.] Bernstein, S., Le problème de l'approximation des fonctions continues sur tout l'axe réel et l'une de ses application, *Bull. Soc. Math. France*, **52** (1924), 399–410.

[9.] Bernstein, S., *Lecons sur les Propriétés Extremales et la Meilleure Approximation des Fonctions Analytiques D'une Variable Réelle*, Gauthier-Villars, Paris, 1926.

[10.] Bernstein, S. N., *Extremal Properties of Polynomials and the Best Approximation of Continuous Functions of a Single Real Variable*, State United Scientific and Technical Publishing House, Moscow, 1937. (In Russian)

[11.] Bernstein, S., Sur les meilleure approximation de $|x|^p$, par des polynômes de degrés trés élevés, *Bull. Acad. Sci. USSR Sér. Math.*, **2** (1938), 181–190.

[12.] Bernstein, S. N., On the best approximation of continuous functions on the whole real axis by means of entire functions of given degree, 3, *Dokl. Akad. Nauk SSSR*, **52** (1946), 565–568. (In Russian)

[13.] Bernstein, S. N., On the best approximation of continuous functions on the whole real axis by means of entire functions of given degree, 5, *Dokl. Akad. Nauk SSSR*, **54** (1946), 479–482. (In Russian)

[14.] Bernstein, S. N., New derivation and generalization of certain formulae of best approximation, *Dokl. Akad. Nauk SSSR*, **54** (1946), 667–668. (In Russian)

[15.] Bernstein, S. N., On limit relations between the constants of the theory of best approximation, *Dokl. Akad. Nauk SSSR*, **57** (1947), 3–5. (In Russian)

[16.] Bernstein, S. N., Limit laws of the theory of best approximation, *Dokl. Akad. Nauk SSSR*, **58** (1947), 525–528. (In Russian)

[17.] Bernstein, S. N., On certain elementary extremal properties of polynomials of several variables, *Dokl. Akad. Nauk SSSR*, **59** (1948), 833–836. (In Russian)

[18.] Bernstein, S. N., On the best approximation of $|x-c|^p$. In: *Collected Works*, Vol. II, pp. 273–280, Akad. Nauk SSSR, Moscow, 1954. (In Russian)

[19.] Boas, R. P., *Entire Functions*, Academic Press, New York, 1954.

[20.] Brychkov, Yu. A. and Prudnikov, A.P., *Integral Transforms of Generalized Functions*, Gordon and Breach Science Publishers, New York, 1989.

[21.] Damelin, S. B., Converse and smoothness theorems for Erdös weights in L_p ($0 < p \leq \infty$), *J. Approx. Theory*, **93** (1998), 349–398.

[22.] Damelin, S. B. and Lubinsky, D. S., Jackson theorems for Erdös weights in L_p ($0 < p \leq \infty$), *J. Approx. Theory*, **94** (1998), 333–382.

[23.] Deift, P., Kriecherbauer, T., McLaughlin, K. T-R., Venakides, S., and Zhou, X., Strong asymptotics of orthogonal polynomials with respect to exponential weights, *Comm. Pure Appl. Math.*, **52** (1999), 1491–1552.

[24.] Demyanov, V. F. and Rubinov, A. M., *Approximate Methods in Optimization Problems*, American Elsevier, New York, 1970.

[25.] DeVore, R. A. and Lorentz, G. G., *Constructive Approximation*, Springer, Berlin, 1993.

[26.] Ditzian, Z. and Lubinsky, D. S., Jackson and smoothness theorems for Freud weights in L_p ($0 < p \leq \infty$), *Constr. Approx.*, **13** (1997), 99–152.

[27.] Ditzian, Z. and Totik, V., *Moduli of Smoothness*, Springer, Berlin, 1987.

[28.] Dzrbasjan, M. M., On metrical criteria of completeness of systems of polynomials in unbounded domains, *Dokl. Akad. Nauk Armenian SSR*, **7** (1947), 3–10. (In Russian)

[29.] Dzrbasjan, M. M., On weighted best approximation of the function $|x|$ on the whole real axis, *Izv. Akad. Nauk Armenian SSR, Ser. Fiz-mat*, **11**(1) (1958), 77–84. (In Russian)

[30.] Dzrbasjan, M. M. and Tavadyan, A. B., On weighted uniform approximation by polynomials of functions of many variables, *Mat. Sb.*, **43**(85) (1957), 227–256. (In Russian)

[31.] Dzyadyk, V. K., On the least upper bounds of best approximations on certain classes of continuous functions, *Dopovidi Akad. Nauk URSR*, Ser. A(7) (1975), 589–592. (In Russian)

[32.] Erdelyi, A., Magnus, W., Oberhettinger, F., and Tricomi, F. G., *Higher Transcendential Functions*, Vol. I, McGraw-Hill, New York, 1953.

[33.] Erdelyi, A., Magnus, W., Oberhettinger, F., and Tricomi, F. G., *Higher Transcendential Functions*, Vol. III, McGraw-Hill, New York, 1955.

[34.] Favard, J., Sur les meilleures procédés d'approximation de certaines classes des fonctions par des polynômes trigonométriques, *Bull. Sci. Math.*, **61** (1937), 209–224, 243–256.

[35.] Fedoryuk, M. V., *Asymptotics: Integrals and Series*, Nauka, Moscow, 1987. (In Russian)

[36.] Freud, G., On Markov-Bernstein type inequalities and their applications, *J. Approx. Theory*, **19** (1977), 22–37.

[37.] Freud, G., Markov-Bernstein type inequalities in $L_p(-\infty, \infty)$. In: Lorentz, G. G., et al. (eds.), *Approximation Theory II*, pp. 369–377, Academic Press, New York, 1976.

[38.] Ganzburg, M. I., The theorems of Jackson and Bernstein in $\mathbf{R}^m$, *Russian Math. Surveys*, **34** (1979), 221–222.

[39.] Ganzburg, M. I., Multidimensional Jackson theorems, *Siberian Math. J.*, **22** (1981), 223–231.

[40.] Ganzburg, M. I., Multidimensional limit theorems of the theory of best polynomial approximations, *Siberian Math. J.*, **23** (1983), 316–331.

[41.] Ganzburg, M. I., Multidimensional Bernstein-type inequalities, *Ukrainian Math. J.*, **34** (1983), 607–610.

[42.] Ganzburg, M. I., On sharp constants for best harmonic approximation of functions of several variables, *Soviet Math. Dokl.*, **37** (1988), 142–145.

[43.] Ganzburg, M. I., Best approximation of sums of elements and a theorem of Newman and Shapiro, *Ukrainian Math. J.*, **41** (1989), 1395–1401.

[44.] Ganzburg, M. I., Limit theorems for the best polynomial approximations in the L_∞ metric, *Ukrainian Math. J.*, **43** (1991), 299–305.

[45.] Ganzburg, M. I., Limit theorems in approximation theory, *Anal. Math.*, **18** (1992), 37–57.

[46.] Ganzburg, M. I., Limit relations in approximation theory and their applications. In: Chui, C. K., Schumaker, L. L. (eds.), *Approximation Theory VIII, Vol 1: Approximation and Interpolation*, pp. 223–232, World Scientific, Singapore, 1995.

[47.] Ganzburg, M. I., Limit theorems and best constants of approximation theory. In: Anastassiou, G. A. (ed.), *Handbook on Analytic-Computational Methods in Applied Mathematics*, pp. 507–569, Chapman & Hall/CRC Press, London/Boca Raton, FL, 2000.

[48.] Ganzburg, M. I., Polynomial inequalities on measurable sets and their applications, *Const. Approx.*, **17** (2001), 275–306.

[49.] Ganzburg, M. I., The Bernstein constant and polynomial interpolation at the Chebyshev nodes, *J. Approx. Theory*, **119** (2002), 193–213.

[50.] Ganzburg, M. I., Limit theorems for polynomial approximations with Hermite and Freud weights. In: Chui, C. K., et al. (eds.), *Approximation Theory X: Abstract and Classical Analysis*, pp. 211–221, Vanderbilt University Press, Nashville, TN, 2002.

[51.] Ganzburg, M. I., Best constants of harmonic approximation on certain classes of differentiable functions, *Bull. London Math. Soc.*, **35** (2003), 355–361.

[52.] Gelfand, I. M. and Shilov, G. E., *Generalized Functions, Vol I: Properties and Operators*, Academic Press, New York, 1964.

[53.] Genchev, T. G., Entire functions of exponential type with polynomial growth on $\mathbf{R}_x^n$. *J. Math. Anal. and Appl.*, **60** (1977), 103–119.

[54.] Gradshteyn, I. S. and Ryzhik, I. M., *Tables of Integrals, Series, and Products*, 5th Edition, Academic Press, San Diego, 1994.

[55.] Gromov, A. Yu., The sharp constant in the Jackson theorem on best approximation by entire functions of exponential type. In: Korneichuk, N. P. (ed.), *Studies in Current Problems of Summation and Approximation of Functions and their Applications*, VI, pp. 210–211, Dniepropetrovsk State University, Dniepropetrovsk, Ukraine, 1975. (In Russian)

[56.] Ibragimov, I. I., On best approximation by polynomials of the functions $[ax + b|x|]|x|^s$ on the interval $[-1, 1]$, *Izv. Akad. Nauk SSSR, Ser. mat.*, **14** (1950), 405–412. (In Russian)

[57.] Kacnel'son, V. E., Equivalent norms in the spaces of entire functions of exponential type, *Math USSR Sbornik*, **21**(1) (1973), 33–55.

[58.] Kadec, M. I., The inverse problem of best approximation of bounded uniformly continuous functions by entire functions of exponential type, and related questions, *Lect. Notes Math.*, **1043** (1984), 591–594.

[59.] Koosis, P., *The Logarithmic Integral II*, Cambridge University Press, Cambridge, UK, 1992.

[60.] Korneichuk, N. P., The best uniform approximation of certain classes of continuous functions, *Dokl. Akad. Nauk SSSR*, **141** (1961), 304–307. (In Russian)

[61.] Korneichuk, N. P., Extremal values of functionals and best approximation on classes of periodic functions, *Izv. Akad. Nauk SSSR*, **35** (1971), 93–124. (In Russian)

[62.] Korneichuk, N. P., *Exact Constants in Approximation Theory*, Cambridge University Press, Cambridge, UK, 1991.

[63.] Krein, M. G., On the best approximation of continuous differentiable functions on the whole real axis, *Dokl. Akad. Nauk SSSR*, **18** (1938), 615–624. (In Russian)

[64.] Kriecherbauer, T. and McLaughlin, K. T-R., Strong asymptotics of polynomials orthogonal with respect to Freud weights, *International Math. Research Notices*, **6** (1999), 299–333.

[65.] Kroó, A. and Szabados, J., Weighted polynomial approximation on the real line, *J. Approx. Theory*, **83** (1995), 41–64.

[66.] Kroó, A. Szabados, J., and Varga, R. S., Weighted polynomial approximation of some entire functions on the real line, *Annals Numer. Math.*, **4** (1997), 405–413.

[67.] Levin, B. Ya., *Distribution of Zeros of Entire Functions*, Amer. Math. Soc., Providence, RI, 1980.

[68.] Levin, B. Ya., Subharmonic majorants and their applications. In: *Theory of Functions of a Complex Variable: Abstracts of Scientific Communications of All-Union Conference*, Kharkov, September 1971, pp. 117–120, Akad. Nauk Ukrainian SSR, Kharkov, 1971. (In Russian)

[69.] Levin, A. L. and Lubinsky, D. S., Christoffel functions, orthogonal polynomials, and Nevai's conjecture for Freud weights, *Constr. Approx.*, **8** (1992), 463–535.

[70.] Levin, A. L. and Lubinsky, D. S., L_p Markov-Bernstein inequalities for Freud weights, *J. Approx. Theory*, **77** (1994), 229–248.

[71.] Levin, A. L. and Lubinsky, D. S., *Christoffel Functions and Orthogonal Polynomials for Exponential Weights on $[-1, 1]$*, Memoirs Amer. Math. Soc., No. 535, **111**, 1994.

[72.] Levin, A. L. and Lubinsky, D. S., *Orthogonal Polynomials for Exponential Weights*, Springer, New York, 2001.

[73.] Logvinenko, V. N. and Sereda, Yu. F., Equivalent norms in the spaces of entire functions of exponential type, *Theory of Function, Functional Analysis, and their Applications*, **20** (1974), 102–111. (In Russian)

[74.] Lorentz, G. G., von Golitschek, M., and Makovoz, Y., *Constructive Approximation: Advanced Problems*, Springer, Berlin, 1996.

[75.] Lubinsky, D. S., *Strong Asymptotics for Extremal Errors and Polynomials Associated with Erdős Type Weights*, Pitman Research Notes in Mathematics, **202**, Longman, Harlow, Essex, 1989.

[76.] Lubinsky, D. S. L_∞ Markov and Bernstein inequalities for Erdős weights, *J. Approx. Theory*, **60** (1990), 188–230.

[77.] Lubinsky, D. S., An update on orthogonal polynomials and weighted approximation on the real line, *Acta Appl. Math.*, **33** (1993), 121–164.

[78.] Lubinsky, D. S., Ideas of weighted polynomial approximation on $(-\infty,\infty)$. In: Chui, C. K., Schumaker, L. L. (eds.), *Approximation Theory VIII.* Vol. 1: *Approximation and Interpolation*, pp. 371–396, World Scientific, Singapore, 1995.

[79.] Lubinsky, D. S., Forward and converse theorems of polynomial approximation for exponential weights on $[-1,1]$, *J. Approx. Theory*, **91** (1997), 1–47, 48–83.

[80.] Lubinsky, D. S., Asymptotics of orthogonal polynomials: Some old, some new, some identities, *Acta Appl. Math.*, **61** (2000), 207–256.

[81.] Lubinsky, D. S., A representation for the Bernstein constant. In: *Advances in Constructive Approximation: Abstracts of Scientific Communications of International Conference*, Nashville, TN, May 14–17, 2003, pp. 31–32, Vanderbilt University, Nashville, TN, 2003.

[82.] Lubinsky, D. S. and Mthembu, T. Z., Orthogonal expansions and the error of weighted polynomial approximation for Erdös weights, *Numer. Funct. Anal. and Optimiz.*, **13** (1992), 327–347.

[83.] Lubinsky, D. S. and Nevai, P., Markov-Bernstein inequalities revisited, *Chinese J. Approx. Theory and its Applications*, **3** (1987), 98–119.

[84.] Lubinsky, D. S. and Saff, E. B., *Strong Asymptotics for Extremal Polynomials Associated with Exponential Weights*, Lect. Notes in Math., **1305**, Springer, Berlin, 1988.

[85.] Malgrange, B., *Ideals of Differentiable Functions*, Oxford University Press, 1966.

[86.] Markushevich, A. I., *Theory of Functions of a Complex Variable*, vol. II, Prentice-Hall, Englewood Cliffs, NJ, 1967.

[87.] Mergelyan, S. N. On weighted approximations by polynomials, *Dokl. Akad. Nauk SSSR*, **97** (1954), 597–600. (In Russian)

[88.] Megelyan, S. N., Weighted approximations by polynomials, *Uspekhi Mat. Nauk*, **11** (1956), 107–152. (In Russian)

[89.] Mhaskar, H. N., Weighted polynomial approximation of entire functions, II, *J. Approx. Theory*, **33** (1981), 59–68.

[90.] Mhaskar, H. N., Weighted polynomial approximation of entire functions, I, *J. Approx. Theory*, **35** (1982), 203–213.

[91.] Mhaskar, H. N., *Introduction to the Theory of Weighted Polynomial Approximation*, World Scientific, Singapore, 1996.

[92.] Mhaskar, H. N., On the degree of approximation in multivariate weighted approximation. In: Buhmann, M. D., Mache, D. M. (eds.), *Advanced Problems in Constructive Approximation*, International Series of Numerical Mathematics, **142**, pp. 129–141, Birkhäuser, Verlag, Basel, 2002.

[93.] Mhaskar, H. N. and Saff, E. B., Extremal problems for polynomials with exponential weights, *Trans. Amer. Math. Soc.*, **285** (1984), 203–234.

[94.] Mhaskar, H. N. and Saff, E. B., Where does the sup-norm of a weighted polynomial live? *Constr. Approx.*, **1** (1985), 71–91.

[95.] Mhaskar, H. N. and Saff, E. B., Where does the L_p-norm of a weighted polynomial live? *Trans. Amer. Math. Soc.*, **303** (1987), 109–124.

[96.] Mthembu, T. Z., Bernstein and Nikolskii inequalities for Erdös weights, *J. Approx. Theory*, **75** (1993), 214–235.

[97.] Nevai, P., Geza Freud, orthogonal polynomials and Christoffel functions: A case study, *J. Approx. Theory*, **48** (1986), 3–167.

[98.] Nevai, P. and Totik, V., Weighted polynomial inequalities, *Constr. Approx.*, **2** (1986), 113–127.

[99.] Nikolskii, S. M., On the best mean approximation by polynomials of the function $|x-c|^s$, *Izv. Akad. Nauk SSSR, Ser. Mat.*, **11** (1947), 139–180. (In Russian)

[100.] Nikolskii, S. M., Inequalities for entire functions of finite degree and their applications in the theory of differentiable functions of many variables, *Trudy Mat. Inst. Steklova*, **38** (1951), 244–278.

[101.] Nikolskii, S. M., *Approximation of Functions of Several Variables and Imbedding Theorems*, Springer, New York-Heidelberg, 1975.

[102.] Plancherel, M. and Polya, G., Fonctions entiéres et intégrales de Fourier multiples. *Comm. Math. Helv.*, **9** (1937), 224–248.

[103.] Pollard, H., Solution of Bernstein's approximation problem, *Proc. Amer. Math. Soc.*, **4** (1953), 869–875.

[104.] Polovina, A. I., On the best uniform approximation by algebraic polynomials of differentiable functions, *Izv. Vysch. Uchebn. Zaved. Mat.*, **13**(12) (1969), 76–82. (In Russian)

[105.] Raitsin, R. A., S. N. Bernstein's limit theorem for best approximations in the mean, and some of its applications, *Izv. Vysch. Uchebn. Zaved. Mat.*, **12**(10) (1968), 81–86. (In Russian)

[106.] Raitsin, R. A., The best mean square approximation, by polynomials and by entire functions of finite degree of functions which have algebraic singular point, *Izv. Vysch. Uchebn. Zaved. Mat.*, **13**(4) (1969), 59–61. (In Russian)

[107.] Rakhmanov, E. A., On asymptotic properties of polynomials orthogonal on the real axis, *Math. USSR Sbornik*, **47** (1984), 155–193.

[108.] Rakhmanov, E. A. Strong asymptotics for orthogonal polynomials associated with exponential weights on **R**. In: Gonchar, A. A., Saff, E. B. (eds.), *Methods in Approximation Theory in Complex Analysis and Mathematical Physics*, pp. 71–97, Nauka, Moscow, 1992.

[109.] Saff, E. B. and Totik, V., *Logarithmic Potentials with External Fields*, Springer, Berlin, 1997.

[110.] Stahl M. and Totik, V., *General Orthogonal Polynomials*, Cambridge University Press, Cambridge, UK, 1992.

[111.] Stein, E. M. and Weiss, G., *Introduction to Fourier Analysis on Euclidean Spaces*, Princeton University Press, Princeton, NJ, 1971.

[112.] Strichartz, R., *A Guide to Distribution Theory and Fourier Transform*, CRC Press, Boca Raton, FL, 1994.

[113.] Szegö, G., *Orthogonal Polynomials*, Colloquium Publications, **23**, Amer. Math. Soc., Providence, RI, 1975.

[114.] Timan, A. F., *Theory of Approximation of Functions of a Real Variable*, MacMillan, New York, 1963.

[115.] Titchmarsh, E. C., *The Theory of Functions*, 2nd edition, Oxford University Press, 1976.

[116.] Totik, V., *Weighted Approximation with Varying Weights*, Lect. Notes in Math., **1569**, Springer, Berlin, 1994.

[117.] Totik, V., Approximation on compact subsets of **R**. In: Buhmann, M. D., Mache, D. H. (eds.), *Advanced Problems in Constructive Approximation, International Series of Numerical Mathematics*, **142**, pp. 263–274, Birkhaüser Verlag, Basel, 2002.

[118.] Van Assche, W., *Asymptotic Properties for Orthogonal Polynomials*, Lect. Notes. in Math., **1265**, Springer, Berlin, 1987.

[119.] Varga, R. S. and Carpenter, A. J., On the Bernstein conjecture in approximation theory, *Constr. Approx.*, **1** (1985), 333–348.

[120.] Vasiliev, R. K., *Chebyshev Polynomials and Approximation Theory on Compact Subsets of the Real Axis*, Saratov University, Saratov, 1998.

[121.] Walsh, J. L. *Interpolation and Approximation by Rational Functions in the Complex Domain*, 5th edition, Colloquium Publications, **20**, Amer. Math. Soc., Providence, RI, 1969.

[122.] Watson, G. N., *A Treatise on the Theory of Bessel Functions*, Cambridge University Press, Cambridge, UK, 1966.

Index

approximation errors
  by weighted polynomials, 8
  by entire functions of exponential type, 8
  by trigonometric polynomials, 96
  multidimensional, 106
asymptotics for
  extremal polynomials, 45
  orthonormal polynomials, 49

Bernstein's approximation problem, 1
Bernstein's inequality, 18

canonical weights, 1, 117
Chebyshev polynomials, 134

elliptic integral, 135
entire function
  of exponential type, 3, 8, 59
  of minimal exponential type, 127
  of order α, 127
Erdös weight, 2, 11, 134
Euler integrals, 128
extremal polynomials
  asymptotics for, 45
  inequalities for, 45, 46
  zeros of, 45

Fourier transform
  of a function, 8, 94
  of a tempered distribution, 8, 90, 92, 95, 114
Freud weight, 2, 11
function
  Airy, 49
  harmonic, 19
  hypergeometric, 117
  λ-homogeneous, 87
  Legendre, 135
  Macdonald, 54
  quasi-increasing, 9
  subharmonic, 37, 57

generalized Paley-Wiener theorem, 91, 114

Joukowski map, 8, 45

Lagrange interpolation, 67, 123, 130
limit theorems, 3, 5, 15, 77
  multidimensional, 105
Legendre polynomials, 134

Markov-Bernstein inequalities, 2
Mhaskar-Rakhmanov-Saff number, 2, 12
Mehler-Heine formula, 100
Mergelyan's method, 127, 128

Nikolskii-type inequalities, 43
normalized extremal polynomials, 45

orthonormal polynomials, 48
  asymptotics for, 49

Paley-Wiener representation, 108
Poisson integral, 19

quasi-norm, 7

sharp constants of approximation, 96
Szegö function, 19
Szegö weight, 2
Sz-Nagy criterion, 94

tempered distribution, 8
trigonometric polynomials, 96

V. A. Markov's inequality, 17
V. A. Markov-type inequality, 18

weight
  canonical, 1, 117
  Erdös, 2, 11, 134
  exponential, 8
  Freud, 2, 13
  Hermite, 117
Whitney theorem, 114

zeros of extremal polynomials, 45

Editorial Information

To be published in the *Memoirs*, a paper must be correct, new, nontrivial, and significant. Further, it must be well written and of interest to a substantial number of mathematicians. Piecemeal results, such as an inconclusive step toward an unproved major theorem or a minor variation on a known result, are in general not acceptable for publication.

Papers appearing in *Memoirs* are generally at least 80 and not more than 200 published pages in length. Papers less than 80 or more than 200 published pages require the approval of the Managing Editor of the Transactions/Memoirs Editorial Board.

As of November 30, 2007, the backlog for this journal was approximately 14 volumes. This estimate is the result of dividing the number of manuscripts for this journal in the Providence office that have not yet gone to the printer on the above date by the average number of monographs per volume over the previous twelve months, reduced by the number of volumes published in four months (the time necessary for preparing a volume for the printer). (There are 6 volumes per year, each usually containing at least 4 numbers.)

A Consent to Publish and Copyright Agreement is required before a paper will be published in the *Memoirs*. After a paper is accepted for publication, the Providence office will send a Consent to Publish and Copyright Agreement to all authors of the paper. By submitting a paper to the *Memoirs*, authors certify that the results have not been submitted to nor are they under consideration for publication by another journal, conference proceedings, or similar publication.

Information for Authors

Memoirs are printed from camera copy fully prepared by the author. This means that the finished book will look exactly like the copy submitted.

Initial submission. The AMS uses Centralized Manuscript Processing for initial submissions. Authors should submit a PDF file using the Initial Manuscript Submission form found at www.ams.org/cgi-bin/peertrack/submission.pl, or send one copy of the manuscript to the following address: Centralized Manuscript Processing, MEMOIRS OF THE AMS, 201 Charles Street, Providence, RI 02904-2294 USA. If a paper copy is being forwarded to the AMS, indicate that it is for it Memoirs and include the name of the corresponding author, contact information such as email address or mailing address, and the name of an appropriate Editor to review the paper (see the list of Editors below).

The paper must contain a *descriptive title* and an *abstract* that summarizes the article in language suitable for workers in the general field (algebra, analysis, etc.). The *descriptive title* should be short, but informative; useless or vague phrases such as "some remarks about" or "concerning" should be avoided. The *abstract* should be at least one complete sentence, and at most 300 words. Included with the footnotes to the paper should be the 2000 *Mathematics Subject Classification* representing the primary and secondary subjects of the article. The classifications are accessible from www.ams.org/msc/. The list of classifications is also available in print starting with the 1999 annual index of *Mathematical Reviews*. The Mathematics Subject Classification footnote may be followed by a list of *key words and phrases* describing the subject matter of the article and taken from it. Journal abbreviations used in bibliographies are listed in the latest *Mathematical Reviews* annual index. The series abbreviations are also accessible from www.ams.org/publications/. To help in preparing and verifying references, the AMS offers MR Lookup, a Reference Tool for Linking, at www.ams.org/mrlookup/.

Electronically prepared manuscripts. The AMS encourages electronically prepared manuscripts, with a strong preference for $\mathcal{A}_{\mathcal{M}}\mathcal{S}$-LaTeX. To this end, the Society has prepared $\mathcal{A}_{\mathcal{M}}\mathcal{S}$-LaTeX author packages for each AMS publication. Author packages include instructions for preparing electronic manuscripts, samples, and a style file that generates the particular design specifications of that publication series. Though $\mathcal{AMS}$-LaTeX is the highly preferred format of TeX, author packages are also available in $\mathcal{AMS}$-TeX.

Authors may retrieve an author package from the AMS website starting from `www.ams.org/tex/` or via FTP to `ftp.ams.org` (login as `anonymous`, enter username as password, and type `cd pub/author-info`). The *AMS Author Handbook* and the *Instruction Manual* are available in PDF format following the author packages link from `www.ams.org/tex/`. The author package can also be obtained free of charge by sending email to `tech-support@ams.org` (Internet) or from the Publication Division, American Mathematical Society, 201 Charles St., Providence, RI 02904-2294, USA. When requesting an author package, please specify $\mathcal{AMS}$-LaTeX or $\mathcal{AMS}$-TeX and the publication in which your paper will appear. Please be sure to include your complete mailing address.

After acceptance. The final version of the electronic file should be sent to the Providence office (this includes any TeX source file, any graphics files, and the DVI or PostScript file) immediately after the paper has been accepted for publication.

Before sending the source file, be sure you have proofread your paper carefully. The files you send must be the EXACT files used to generate the proof copy that was accepted for publication. For all publications, authors are required to send a printed copy of their paper, which exactly matches the copy approved for publication, along with any graphics that will appear in the paper.

Accepted electronically prepared files can be submitted via the web at `www.ams.org/submit-book-journal/`, sent via FTP, or sent on CD-Rom or diskette to the Electronic Prepress Department, American Mathematical Society, 201 Charles Street, Providence, RI 02904-2294 USA. TeX source files, DVI files, and PostScript files can be transferred over the Internet by FTP to the Internet node `ftp.ams.org` (130.44.1.100). When sending a manuscript electronically via CD-Rom or diskette, please be sure to include a message identifying the paper as a Memoir.

Electronically prepared manuscripts can also be sent via email to `pub-submit@ams.org` (Internet). In order to send files via email, they must be encoded properly. (DVI files are binary and PostScript files tend to be very large.)

Electronic graphics. Comprehensive instructions on preparing graphics are available at `www.ams.org/jourhtml/`. A few of the major requirements are given here.

Submit files for graphics as EPS (Encapsulated PostScript) files. This includes graphics originated via a graphics application as well as scanned photographs or other computer-generated images. If this is not possible, TIFF files are acceptable as long as they can be opened in Adobe Photoshop or Illustrator. No matter what method was used to produce the graphic, it is necessary to provide a paper copy to the AMS.

Authors using graphics packages for the creation of electronic art should also avoid the use of any lines thinner than 0.5 points in width. Many graphics packages allow the user to specify a "hairline" for a very thin line. Hairlines often look acceptable when proofed on a typical laser printer. However, when produced on a high-resolution laser imagesetter, hairlines become nearly invisible and will be lost entirely in the final printing process.

Screens should be set to values between 15% and 85%. Screens which fall outside of this range are too light or too dark to print correctly. Variations of screens within a graphic should be no less than 10%.

Inquiries. Any inquiries concerning a paper that has been accepted for publication should be sent to `memo-query@ams.org` or directly to the Electronic Prepress Department, American Mathematical Society, 201 Charles St., Providence, RI 02904-2294 USA.

Editors

This journal is designed particularly for long research papers, normally at least 80 pages in length, and groups of cognate papers in pure and applied mathematics. Papers intended for publication in the *Memoirs* should be addressed to one of the following editors. The AMS uses Centralized Manuscript Processing for initial submissions to AMS journals. Authors should follow instructions listed on the Initial Submission page found at www.ams.org/memo/memosubmit.html.

Algebra to ALEXANDER KLESHCHEV, Department of Mathematics, University of Oregon, Eugene, OR 97403-1222; email: ams@noether.uoregon.edu

Algebraic geometry and its application to MINA TEICHER, Emmy Noether Research Institute for Mathematics, Bar-Ilan University, Ramat-Gan 52900, Israel; email: teicher@macs.biu.ac.il

Algebraic geometry to DAN ABRAMOVICH, Department of Mathematics, Brown University, Box 1917, Providence, RI 02912; email: amsedit@math.brown.edu

Algebraic number theory to V. KUMAR MURTY, Department of Mathematics, University of Toronto, 100 St. George Street, Toronto, ON M5S 1A1, Canada; email: murty@math.toronto.edu

Algebraic topology to ALEJANDRO ADEM, Department of Mathematics, University of British Columbia, Room 121, 1984 Mathematics Road, Vancouver, British Columbia, Canada V6T 1Z2; email: adem@math.ubc.ca

Combinatorics to JOHN R. STEMBRIDGE, Department of Mathematics, University of Michigan, Ann Arbor, Michigan 48109-1109; email: FRS@umich.edu

Complex analysis and harmonic analysis to ALEXANDER NAGEL, Department of Mathematics, University of Wisconsin, 480 Lincoln Drive, Madison, WI 53706-1313; email: nagel@math.wisc.edu

Differential geometry and global analysis to LISA C. JEFFREY, Department of Mathematics, University of Toronto, 100 St. George St., Toronto, ON Canada M5S 3G3; email: jeffrey@math.toronto.edu

Functional analysis and operator algebras to DIMITRI SHLYAKHTENKO, Department of Mathematics, University of California, Los Angeles, CA 90095; email: shlyakht@math.ucla.edu

Geometric analysis to WILLIAM P. MINICOZZI II, Department of Mathematics, Johns Hopkins University, 3400 N. Charles St., Baltimore, MD 21218; email: trans@math.jhu.edu

Geometric analysis to MARK FEIGHN, Math Department, Rutgers University, Newark, NJ 07102; email: feighn@andromeda.rutgers.edu

Harmonic analysis, representation theory, and Lie theory to ROBERT J. STANTON, Department of Mathematics, The Ohio State University, 231 West 18th Avenue, Columbus, OH 43210-1174; email: stanton@math.ohio-state.edu

Logic to STEFFEN LEMPP, Department of Mathematics, University of Wisconsin, 480 Lincoln Drive, Madison, Wisconsin 53706-1388; email: lempp@math.wisc.edu

Number theory to JONATHAN ROGAWSKI, Department of Mathematics, University of California, Los Angeles, CA 90095; email: jonr@math.ucla.edu

Partial differential equations to GUSTAVO PONCE, Department of Mathematics, South Hall, Room 6607, University of California, Santa Barbara, CA 93106; email: ponce@math.ucsb.edu

Partial differential equations and dynamical systems to PETER POLACIK, School of Mathematics, University of Minnesota, Minneapolis, MN 55455; email: polacik@math.umn.edu

Probability and statistics to RICHARD BASS, Department of Mathematics, University of Connecticut, Storrs, CT 06269-3009; email: bass@math.uconn.edu

Real analysis and partial differential equations to DANIEL TATARU, Department of Mathematics, University of California, Berkeley, Berkeley, CA 94720; email: tataru@math.berkeley.edu

All other communications to the editors should be addressed to the Managing Editor, ROBERT GURALNICK, Department of Mathematics, University of Southern California, Los Angeles, CA 90089-1113; email: guralnic@math.usc.edu.

Titles in This Series

900 **Wolfgang Bertram,** Differential geometry, Lie groups and symmetric spaces over general base fields and rings, 2008

899 **Piotr Hajłasz, Tadeusz Iwaniec, Jan Malý, and Jani Onninen,** Weakly differentiable mappings between manifolds, 2008

898 **John Rognes,** Galois extensions of structured ring spectra/Stably dualizable groups, 2008

897 **Michael I. Ganzburg,** Limit theorems of polynomial approximation with exponential weights, 2008

896 **Michael Kapovich, Bernhard Leeb, and John J. Millson,** The generalized triangle inequalities in symmetric spaces and buildings with applications to algebra, 2008

895 **Steffen Roch,** Finite sections of band-dominated operators, 2008

894 **Martin Dindoš,** Hardy spaces and potential theory on C^1 domains in Riemannian manifolds, 2008

893 **Tadeusz Iwaniec and Gaven Martin,** The Beltrami Equation, 2008

892 **Jim Agler, John Harland, and Benjamin J. Raphael,** Classical function theory, operator dilation theory, and machine computation on multiply-connected domains, 2008

891 **John H. Hubbard and Peter Papadopol,** Newton's method applied to two quadratic equations in $\mathbb{C}^2$ viewed as a global dynamical system, 2008

890 **Steven Dale Cutkosky,** Toroidalization of dominant morphisms of 3-folds, 2007

889 **Michael Sever,** Distribution solutions of nonlinear systems of conservation laws, 2007

888 **Roger Chalkley,** Basic global relative invariants for nonlinear differential equations, 2007

887 **Charlotte Wahl,** Noncommutative Maslov index and eta-forms, 2007

886 **Robert M. Guralnick and John Shareshian,** Symmetric and alternating groups as monodromy groups of Riemann surfaces I: Generic covers and covers with many branch points, 2007

885 **Jae Choon Cha,** The structure of the rational concordance group of knots, 2007

884 **Dan Haran, Moshe Jarden, and Florian Pop,** Projective group structures as absolute Galois structures with block approximation, 2007

883 **Apostolos Beligiannis and Idun Reiten,** Homological and homotopical aspects of torsion theories, 2007

882 **Lars Inge Hedberg and Yuri Netrusov,** An axiomatic approach to function spaces, spec tral synthesis and Luzin approximation, 2007

881 **Tao Mei,** Operator valued Hardy spaces, 2007

880 **Bruce C. Berndt, Geumlan Choi, Youn-Seo Choi, Heekyoung Hahn, Boon Pin Yeap, Ae Ja Yee, Hamza Yesilyurt, and Jinhee Yi,** Ramanujan's forty identities for Rogers-Ramanujan functions, 2007

879 **O. García-Prada, P. B. Gothen, and V. Muñoz,** Betti numbers of the moduli space of rank 3 parabolic Higgs bundles, 2007

878 **Alessandra Celletti and Luigi Chierchia,** KAM stability and celestial mechanics, 2007

877 **María J. Carro, José A. Raposo, and Javier Soria,** Recent developments in the theory of Lorentz spaces and weighted inequalities, 2007

876 **Gabriel Debs and Jean Saint Raymond,** Borel liftings of Borel sets: Some decidable and undecidable statements, 2007

875 **C. Krattenthaler and T. Rivoal,** Hypergéométrie et fonction zêta de Riemann, 2007

874 **Sonia Natale,** Semisolvability of semisimple Hopf algebras of low dimension, 2007

873 **A. J. Duncan,** Exponential genus problems in one-relator products of groups, 2007

872 **Anthony V. Geramita, Tadahito Harima, Juan C. Migliore, and Yong Su Shin,** The Hilbert function of a level algebra, 2007

871 **Pascal Auscher,** On necessary and sufficient conditions for L^p-estimates of Riesz transforms associated to elliptic operators on $\mathbb{R}^n$ and related estimates, 2007

TITLES IN THIS SERIES

- 870 **Takuro Mochizuki,** Asymptotic behaviour of tame harmonic bundles and an application to pure twistor D-modules, Part 2, 2007
- 869 **Takuro Mochizuki,** Asymptotic behaviour of tame harmonic bundles and an application to pure twistor D-modules, Part 1, 2007
- 868 **Gelu Popescu,** Entropy and multivariable interpolation, 2006
- 867 **Vilmos Totik,** Metric properties of harmonic measures, 2006
- 866 **William Craig,** Semigroups underlying first-order logic, 2006
- 865 **Nathanial P. Brown,** Invariant means and finite representation theory of $C*$-algebras, 2006
- 864 **John M. Lee,** Fredholm operators and Einstein metrics on conformally compact manifolds, 2006
- 863 **M. Lübke and A. Teleman,** The Universal Kobayashi-Hitchin correspondence on Hermitian manifolds, 2006
- 862 **Alberto Canonaco,** The Beilinson complex and canonical rings of irregular surfaces, 2006
- 861 **Leon A. Takhtajan and Lee-Peng Teo,** Weil-Petersson metric on the universal Teichmüller space, 2006
- 860 **Thomas M. Fiore,** Pseudo limits, biadjoints and pseudo algebras: Categorical foundations of conformal field theory, 2006
- 859 **N. Arcozzi, R. Rochberg, and E. Sawyer,** Carleson measures and interpolating sequences for Besov spaces on complex balls, 2006
- 858 **Enrico Valdinoci, Berardino Sciunzi, and Vasile Ovidiu Savin,** Flat level set regularity of p-Laplace phase transitions, 2006
- 857 **Donatella Danielli, Nocola Garofalo, and Duy-Minh Nhieu,** Non-doubling Ahlfors measures, perimeter measures, and the characterization of the trace spaces of Sobolev functions in Carnot-Carathéodory spaces, 2006
- 856 **Vladimir Bolotnikov and Harry Dym,** On boundary interpolation for matrix valued Schur functions, 2006
- 855 **Yevgenia Kashina, Yorck Sommerhäuser, and Yongchang Zhu,** On higher Frobenius-Schur indicators, 2006
- 854 **Noam Greenberg,** The role of true finiteness in the admissible recursively enumerable degrees, 2006
- 853 **Joachim Krieger,** Stability of spherically symmetric wave maps, 2006
- 852 **Viorel Barbu, Irena Lasiecka, and Roberto Triggiani,** Tangential boundary stabilization of Navier-Stokes equations, 2006
- 851 **Jie Wu,** On maps from loop suspensions to loop spaces and the shuffle relations on the Cohen groups, 2006
- 850 **Siegfried Echterhoff, S. Kaliszewski, John Quigg, and Iain Raeburn,** A categorical approach to imprimitivity theorems for C^*-dynamical systems, 2006
- 849 **Katsuhiko Kuribayashi, Mamoru Mimura, and Tetsu Nishimoto,** Twisted tensor products related to the cohomology of the classifying spaces of loop groups, 2006
- 848 **Bob Oliver,** Equivalences of classifying spaces completed at the prime two, 2006
- 847 **Eric T. Sawyer and Richard L. Wheeden,** Hölder continuity of weak solutions to subelliptic equations with rough coefficients, 2006
- 846 **Victor Beresnevich, Detta Dickinson, and Sanju Velani,** Measure theoretic laws for lim–sup sets, 2006

For a complete list of titles in this series, visit the
AMS Bookstore at **www.ams.org/bookstore/**.